Dogs and Their Humans in Pacific Island Interspecies Cultures

Dogs and Their Humans in Pacific Island Interspecies Cultures explores diverse canine–human relationships in ancient and contemporary Polynesia, Micronesia, Melanesia, and Australia. Anthropologically combining ethnographic, archaeological, biological, and linguistic evidence, the chapters demonstrate that from Hawaiian backyards to the Australian outback, both dogs and humans act differently depending on the cultural conditions connecting them in their multispecies communities.

This book rewards readers with a profound understanding of the intricate and dynamic relationships between dogs and humans, highlighting the significant impact these interspecies interactions have on shaping cultural and ecological landscapes that affect empathy, customs, genes, and ecosystems.

Roger Ivar Lohmann is Associate Professor of Anthropology at Trent University, Oshawa, Ontario, Canada. His interspecies household in Markham includes a human called Heather M.-L. Miller and two rescue Shiba Inus: Wasabi (a would-have-been meat dog from a slaughterhouse near Beijing, China) and Shoga (a former puppy mill mother abandoned on the streets of Scarborough, Ontario).

"This wonderfully wide-ranging and multifaceted volume offers a long overdue focus on the inter-species culture (felicitous phase) of dogs and people throughout the Pacific. It roams across geography and through archeology and anthropology to explore the rich lives of dogs alongside people in the Pacific lands. There is also fresh light and some deep thoughts about shared human–canine lives that will be interesting and useful for those interested in people and their dogs wherever they are found."

Clive Wynne, *Professor of Psychology, Arizona State University, Author of* Dog Is Love: Why and How Your Dog Loves You

"This groundbreaking book explores complex relationships between humans and dogs through the innovative lens of interspecies culture, illustrating how people and dogs culturally shape each other. Contributors illuminate worlds that dogs and people have built together—sometimes with affection, sometimes with ambivalence, sometimes through absence or with violence—but always in relationship. This first comprehensive anthropological study of dogs across the Pacific is essential reading for multispecies researchers and all those who love dogs and want to know more about the innumerable and fascinating ways our lives are entangled."

Yasmine Musharbash, *Associate Professor in the School of Archaeology and Anthropology, Australian National University*

"Across Oceania dogs have held, and continue to hold, a diversity of culturally informed and environmentally influenced roles ranging from family members to food. Yet, anthropological forays into the complex interspecies relationships between Dogs and their Humans are rare. In this much anticipated volume, Lohmann brings together a diverse group of anthropologists to discuss the intertwined lives of dogs and humans. The eloquently written and highly accessible case studies within are certain to appeal to researchers and dog lovers alike."

Justin Cramb, *Assistant Professor of Archaeology, University of Alaska Fairbanks*

"Dogs and Their Humans in Pacific Island Interspecies Cultures is an important contribution to the burgeoning literature about human–canine relationships. Detailed case studies bring the myriad and complex ways our two species interact to light. Finely grained and nuanced observations and analyses of canine cultures (feral, stray, pampered companion, hunter, guard, food) and human cultures (classes, ethnicities, genders) show how dogs and humans shape each other while creating diverse human–canine interspecies cultures."

Bryan Cummins, *Author of* Our Debt to the Dog: How the Domestic Dog Helped Shape Human Societies

Dogs and Their Humans in Pacific Island Interspecies Cultures

Edited by
Roger Ivar Lohmann

Routledge
Taylor & Francis Group
NEW YORK AND LONDON

First published 2026
by Routledge
605 Third Avenue, New York, NY 10158

and by Routledge
4 Park Square, Milton Park, Abingdon, Oxon OX14 4RN

Routledge is an imprint of the Taylor & Francis Group, an informa business

ISBN: 978-1-041-11181-8 (hbk)
ISBN: 978-1-041-11182-5 (pbk)
ISBN: 978-1-003-65869-6 (ebk)

DOI: 10.4324/9781003658696

Typeset in Times New Roman
by Taylor & Francis Books

Contents

Illustrations

Figures

Maps

Table

Contributors

Laurence Marshall Carucci, Letters and Science Distinguished Professor Emeritus of Anthropology at Montana State University, holds a Ph.D. in Anthropology from The University of Chicago. Since 1976 he conducted over seven years of fieldwork with Marshall Islanders. His research concerns cultural change among Enewetak/Ujelang people and others who endured World War II and subsequent United States nuclear tests in Marshall Islands.

Lise M. Dobrin (Ph.D. 1999 in Linguistics from the University of Chicago) is Professor in the University of Virginia Department of Anthropology and Linguistics Program. Her 2012 book *Concreteness in Grammar: The Noun Class Systems of the Arapesh Languages* is based on field research in Wautogik village, Papua New Guinea. Dobrin has published on endangered language linguistics, field methods, ethics, language shift, and Arapesh ethnography.

Richard Feinberg (Ph.D. University of Chicago, 1974) is Professor Emeritus of Anthropology at Kent State University. He conducted fieldwork with Native North Americans and Polynesians in the southwestern Pacific. In 2016 he was elected honorary fellow of the Association for Social Anthropology in Oceania. In 2019 he was Fulbright Distinguished Chair of Anthropology at Palacký University, Olomouc, Czech Republic. He is president-elect of the Association of Senior Anthropologists.

Karen Greig is Senior Lecturer in the Archaeology Programme at Ōtākou Whakaihu Waka University of Otago and Co-Director of Southern Pacific Archaeological Research in Aotearoa New Zealand. Her 2017 Ph.D. combined archaeology, anthropology, and ancient DNA studies to investigate the introduction and dispersal of dogs throughout Oceania. Her multi-species household includes her human family members, two dogs, and a cat.

Loukas Koungoulos is a postdoctoral scholar in Archaeology at the University of Western Australia, based at the Australian Museum. His 2022 Ph.D. focused on the history, archaeology, and evolution of the dingo. Loukas's work examines the deep history of animal–human interactions in the Asia-Pacific region, with a focus on the origins and dispersal of domestic canids and the rewilding of their descendants.

Roger Ivar Lohmann (Ph.D. 2000, University of Wisconsin–Madison) is Associate Professor of Anthropology at Trent University. His interspecies household in Markham, Ontario, Canada includes a human called Heather M.-L. Miller and two rescue Shiba Inus, named Wasabi (a would-have-been meat dog from a slaughterhouse near Beijing, China) and Shoga (a former puppy mill mother abandoned on the streets of Scarborough, Ontario).

Lynn Morrison is Professor of Anthropology at the University of Hawai'i at Hilo. Her research includes cross-cultural analyses of women's risk for HIV/AIDS and immune and stress response using surveys, diaries, and biomarkers. Assessing effects of natural disasters on people and animals led her to the intersection of animal welfare, human health, and cultural attitudes toward dogs. Lynn is a founder of Action 4 Animals Hawaii.

Jordan Prokosch (Ph.D, Washington State University, 2021) has 12 years of experience working with Marshall Islanders. He has lived in Majuro as a high school teacher, college instructor, consultant, and ethnographic researcher. His research focuses on migration, culture change, hybridity, and transnational social networks in the post-colonial context. His most recent work in Majuro contributed to a local government youth skill development project.

Cathleen Pyrek, a lifelong lover of dogs, holds an M.A. in anthropology from Kent State University, where she also taught anthropology. She has enjoyed collaborating with noted anthropologist Rick Feinberg following field research in Samoa and Tokelau. She currently lives in Louisville, Kentucky with her husband, Patrick and their dog, Cooper.

Micah Van der Ryn holds a Master's in visual anthropology from University of Southern California and a Ph.D. in social anthropology from University of Auckland. From 1997–2019 he worked as researcher, filmmaker, and educator at American Samoa Community College. He is married to Faaiuga Faaofonuu with whom he has four children. He received the Samoan matai (chieftain) name of Fepulea'i in 2004.

1 Human–Canine Interspecies Cultures in Oceania and in General

An Introduction

Roger Ivar Lohmann

Introduction

"Man's best friend" both captures and neglects much about dogs' varied relationships with humans in culturally diverse communities. Humans may learn in their societies to treat dogs as companions, helpmeets, inferiors, or livestock. Dogs take the cultural cues and constraints on offer and respond accordingly. Canines trained with specialized skills—be they hunting, guarding, shepherding, detecting, or entertaining—are vital participants in distinct cultural endeavors with their humans. Dogs routinely offered food will grow plump and become sleek tail-waggers around their people. Those used to being attacked with thrown stones will become injured, skittish, or threatening. Depending on the cultural setting, dogs may be free to come and go, be leashed, or be caged. Their lives, routines, and skills are different as a result.

Humans encountering dogs who were trained up in their shared cultural environment will find them behaving in ways that seem to confirm their expectations of what dogs are like. When we encounter dogs who have learned other ways of acting, we realize that generalizing about dogs, like people, needs to take cultural variation into account. Various forms of kindness, cruelty, and neutrality toward dogs are known and learned as normalized practice in different societies and in different sectors of the same societies. Such norms involve and effect both dogs and people. Dogs' and their humans' habits, health, home ranges, and even genetic profiles change in diverse ways as a result of their interspecies histories in culturally different societies.

Attitudes toward one another as friend, fool, or fiend are merely surface ripples on a deep ocean of varied and intricate human–canine interaction styles and their consequences. These variations are social conventions to which we—humans and dogs participating in common communities—train one another. In this volume, we sample the range and depth of this quintessential trans-species partnership in societies across the Pacific Ocean and the lands it surrounds.

DOI: 10.4324/9781003658696-1

Figure 1.1 An Asabano man named Yalowad chose to pose for his picture with a hunting dog and his bow and arrows in Yakob village, Duranmin, Papua New Guinea, 2007. Photo: Roger Lohmann.

In diverse societies of Australia, Melanesia, Micronesia, and Polynesia, dogs and humans have developed a wide range of attitudinal and interactive forms. Humans' and dogs' understandings and ecological engagements differ depending on the traditions they enact with one another. The effects show even on canine displays of cleverness. Dogs were made dull as caged livestock in ancient Hawai'i (Williams et al. 2018). They were made sharp, independent hunting partners in Australia (Koungoulos and Fillios 2020). People whose dogs are pets consequently have mindsets and impacts that differ from those of people who customarily avoid canines or practice other sorts of partnerships with them. The difference between regarding a dog as a beloved family member, as a resource to be slaughtered for meat, or as a vital assistant in the food quest is palpable. And these options far from exhaust the diversity of canine–human cultural engagements.

People are influenced by canine skills, behaviors, and physiques that themselves resulted from interactions with humans who were enculturated with particular beliefs and skills. The cultural influences connecting dogs and their humans are circular rather than one-way. Both are active and responsive to one another in their own ways. Each has their part to play in diverse cultural dramas that involve both species.

Around the world, humans and dogs have joined in common societies and sorted into networks of relative roles and statuses. We share resources and dangers, but also compete, quarrel, and kill one another. Together, we achieve feats that would be impossible for either species on its own. In Tasmania, the first British colonists avoided starvation by cooperatively hunting with the dogs they introduced there for the first time in the early 19th century (Boyce 2006). In the Canadian Arctic, Inuit people and Qimmiq dogs have cooperated for two millennia to locate seals by canine smelling, kill them by human harpooning, and transport them by dog-powered sledding, thereby surviving sustainably on winter ice sheets (Ameen et al. 2019). Dogs and humans have provided mutual protection, shared in work, and protected, charmed, and amused each other. But there are exploitations and terrible abuses as well as mutual support, from careless abandonment to torture of dogs by humans in their societies (Dugnoille 2019). As Bryan Cummins (2013:ix) observes, humankind owes dogkind both gratitude and an apology, not to mention decent treatment and policies that take dogs' perspectives and our common wellbeing to heart. I hope readers of this volume will be moved to enact and promote these where they live.

In what follows, I summarize current scientific knowledge about the origin and spread of dogs to diverse locales in Oceania. I point to the cultural practices and capabilities of human–canine interactions that generated these developments. I then explore the cultural aspects of dog–human engagement under the rubric of human–canine interspecies culture. With this term I identify forms and effects of socially learned information that require the participation of both people and dogs. I next summarize the origins, growth, limits, and potential of the interspecies culture concept more broadly. Summaries of this volume's chapters are presented next, highlighting what I see as examples of human–canine interspecies culture in each. My concluding thesis is that diverse dog–human engagements and their resultant customs, behaviors, and physical consequences are cultural co-creations of generations of both species.

Dog Origins and Spread to Oceania

Dogs derive from Eurasian gray wolf ancestors and are famously regarded as the first species to be domesticated in relationship with humans (Larson et al. 2012). Wolves and humans both rely on membership and communication within social groups to secure food and protection, suggesting parallels in our evolution even before domesticated dogs appeared (Hall and Sharp 1978). Although "domestication" is not precisely or consistently defined, it generally entails significant genetic changes in a population or species caused by breeding selection. Sometimes domestication is supposed to be driven within a single species' evolving lineage, as when "cultural moderation of breeding patterns" is suggested to have caused the evolution of gracility in humans through self-domestication (Bendarik 2008:12). Usually, however, domesticating selection is understood to derive from a mutually advantageous association with another species. Humans' intellectual and cultural adaptations can provide opportunities, help, or direction that change the evolutionary pressures for members of another species who are constrained by and adapt to these conditions. In this way, "domesticated" versions of wolves—dogs—evidently diverged from their "wild" cousins and have since continued to evolve through distinctive selective pressures into their diverse breeds.

Melinda Zeder considers the influence of culture (especially human culture) to be what distinguishes domestication from other forms of symbiosis or mutualism: "This capacity for the cultural transmission of learned behavior", Zeder (2012:228) observes, "ramps up the mutualism between humans and emerging domesticates". For example, meat-eating cultures whose human carriers discarded bones and scraps could have created a situation favoring the self-selection of wolves who were less fearful of humans and willing to stay closer to this food source than other wolves, leading to the evolution of more dog-like descendants (Coppinger and Coppinger 2002). Wolves' exceptional "cultural memory" is one of the characteristics that would have aided their domestication into dogs as generations of people selectively fed the wolves who harmonized better with humans (Mech and Janssens 2021:304). Similarly, people whose cultures came to include adopting wolf puppies could have favored greater reproductive success of friendlier or more communicatively competent wolf-dogs in their company (Hare et al. 2002).

Experimental selection for friendliness toward humans in foxes generated substantial success in fewer than 20 generations and also gave rise to a suite of associated anatomical traits found among some dogs, such as floppy ears, patchy coats, and curly tails (Dugatkin and Trut 2017). These domesticated foxes' distinct, genetically based traits can be traced back to known, cultural interventions by particular humans with distinct, socially learned perspectives, abilities, and goals. Artificial selection expresses cultural preferences in altered dog gene pools, genotypes, and phenotypes. This has been demonstrated in changes to canid bodies, behavioral tendencies, and even abilities to take on human perspectives (Kaminski et al. 2009; Larson et al. 2012).

While humans are often thought of as the active parties in domestication, members of any or all partner species may set in place the characteristic selective pressures (Stahl 2016:61). Both canines and humans can take the initiative in seeking out the advantages of living close to one another. The association changes the selective pressures and opportunities for both species. Alan Beck's (2002 [1973]:xv) groundbreaking study of urban stray dogs in Baltimore credits humankind as the creator of dogkind, while acknowledging that the influence has gone both ways and regarding both species as domesticated.

Dog–human partnerships of various kinds can increase potential survival, reproduction, and ecological flexibility for members of both species who embark upon it as a part of shared interspecies cultures (Shipman 2021; for plants and people, cf. van der Veen 2014:800). Dogs and humans have co-evolved both genetically and culturally (Schleidt and Shalter 2003). The diverse preferences and technologies of ethnic groups who (often unwittingly) influence another species' patterns of mate and trait selection over many generations are one factor of domestication; so are the individual motives and agency of other sentient species who are in a domesticating relationship. They often have a choice whether to continue the association, and what that association will be. Moreover, the example of dogs illustrates that domestication is an ongoing process that can be facilitated by care, taming, and commensality (Losey 2021). Not just genetic changes, but also socially learned "enskillment" of both participating species for their joint undertakings plays an important part in the domestication partnership between humans and dogs (Losey et al. 2020:198).

Genetic and morphological indicators of canid domestication extend more than 33 millennia into the past in a Siberian case (Druzhkova et al. 2013). An Upper Paleolithic site in the Czech Republic yielded three skulls identified as dogs, based on their short snouts and skull lengths and wide palates and brain cases compared to wolves'. Moreover, one of the skulls was buried with a mammoth bone placed in its mouth (Germonpré et al. 2012). This suggests social bonding and commensal partnership between human groups and the earliest wolf-dogs with whom they lived. Another indicator of the antiquity of culturally shaped human–dog cooperation is a 14,000-year-old burial of a young dog in Germany who suffered from distemper and whose survival points to care by human associates (Janssens et al. 2018).

After dogs first appeared in Eurasia, they expanded in the company of their humans around the world (Perri et al. 2021). They crossed ethnic barriers to join new human societies through trade or capture. Sometimes they left particular people—or all people—behind to found fully feral dog societies like those of wild New Guinea highland singing dogs (Surbakti et al. 2020), who sometimes interbred with dogs attached to culturally diverse human villages (Dwyer and Minnegal 2021). In Southeast Asia and Oceania, canids accompanied their humans beyond the range of their gray wolf cousins, thereby at least temporarily halting the gene flow that occasionally occurred elsewhere (Greig et al. 2015b:466). Dogs are found in places that neither they nor their wolf ancestors could have inhabited without human assistance—and vice versa (Qikiqtani Inuit Association 2014:9). Dogs encountered not only different natural selection pressures, but also different breeding, treatment, and training depending on the countries to which people imported them, and hence the cultures under whose influence they came. In this way canids diversified not just in their distribution, but also in their shape and behavioral habits.

Loukas Koungoulos, Jane Balme, and Sue O'Connor (2023:2) write that the first canids to arrive in Oceania are the ancestors of Australia's dingoes and the very closely related New Guinea singing dogs (so called because of their propensity to howl rather than bark) "which were almost certainly introduced to New Guinea in a tamed or domesticated state at approximately the same time". Their earliest direct archaeological traces, dating to around 3,300 years ago in both Australia and islands to the north, indicate that transport from Asia would have required sea crossings with humans on watercraft (Balme and O'Connor 2021). Dogs dispersed across much of Australia, Melanesia, Micronesia, and Polynesia, first arriving in eastern Polynesia by the 14th century CE (Cramb 2022; Greig et al. 2018; Koungoulos and Fillios 2020:3).

Map 1.1 Oceania. Map: Les O'Neill.

In some places, like Australia and parts of Melanesia, the humans who had first discovered Oceania via Southeast Asia were there tens of thousands of years before dogs arrived (Manne et al. 2020). On diverse islands in Micronesia (Wickler 2004:33–34) and Polynesia, including Aotearoa New Zealand (Greig et al. 2015a), and the Cook Islands (Cramb and Hadden 2020), Polynesian dogs and people first colonized lands as a team. Many islands became populated with "native" dogs often described as small, bandy-legged canines (Titcomb 1969). Europeans characterized them as unintelligent, fearful, and prone neither to barking nor howling; all of these traits are attributable to socialization among local and contemporary humans and other dogs as puppies—in other words, they are at least partly expressions of the cultures that these dogs participated in (Williams et al. 2018:3).

Other islands remained devoid of canines or returned to doglessness after trial introductions. Elsewhere still, canids were first introduced to already peopled islands by European explorers and colonists. In Tasmania, for example, British dogs were the first canids to settle. They joined Aboriginal societies with evident mutual enthusiasm (Jones 1970). Other examples of such late introductions include Captain Cook's 18th century canine gifts in New Caledonia (Lyons 1986:3–4). European dogs often participated in their humans' first contacts with Indigenous cultures in New Guinea and elsewhere (Golub 2005). Just like people, successive migrant dogs have often mixed with or replaced genetically and culturally distinct populations of their own species whose ancestors had arrived earlier.

Dog origins, distribution, and variants in body and behavior express distinct, internally variable, and changing cultures. Dogs would not have diverged from wolves without invented associational options becoming widespread and transgenerational through social learning among both human and canine interactors. Dogs could not have come to Oceania without cultural ideas like navigation skills, cultural behaviors like care, or cultural artifacts like watercraft. Dogs of particular and varying selected and maintained breeds and training traditions have colonized the Pacific Islands in joint ventures with humans in ancient Polynesian double-hulled canoes, 18th-century British sailing ships, and in the holds of jets. Their very presence there is an expression of culture crossing species lines.

Human–Canine Interspecies Culture

Dogs and their humans, by grouping and interacting according to diverse socially learned conventions, have altered one another's forms, acts, distributions, and things in ways that neither species could have done on its own. The species *Canis familiaris* itself (to which dogs have been generally assigned), the diversity of canid types and breeds, and the mere presence of dogs in Oceania are all effects of socially learned information traditions necessarily enacted by the engagement of people and dogs. In other words, these phenomena are human–canine, interspecies, and cultural. Genetic evolution and ecological processes alone were insufficient to generate these biological outcomes. Socially learned ideals, abilities, and goals—that is, cultural inputs—were necessary. Humans could not have achieved these results without dogs, nor dogs without humans. As such, they are manifestations of human–canine interspecies culture: socially learned information from traditions that are integrated, variegated, changeable, and diverse, including its material, biological, behavioral, and mental effects.

People may learn in their societies to routinely belittle dogs, to eat them, to beat them, or to coddle them. In some societies but not others, dogs are employed as guards, herders, or hunters. Many details of possible dog–human engagement are invented, shared, stabilize as traditions, and diffuse to other societies. They change over time and vary across social and ethnic groupings in different regions. Depending on which cultural notions are local and current, dogs may be regarded and treated as vermin, as providers, as spirits, as dangerous, as dirty, as commodities, as irrelevant, or as family. Depending on culture, the same engagements with dogs, such as neutering or movement control, may be regarded as kind or cruel, and household dogs may be favored or tabooed (Orr 2016).

Subject to cultural invention, inertia, and variation, a society's dogs may be outfitted with technologies and products designed just for them and their roles vis-à-vis their humans: dogsleds, carts, leashes, vaccines, special foods, or none of the above. They may be cherished or left to themselves; mistreated or discarded. They may or may not be subjected to grooming, docking, or infanticide. Cultural practice in a society may be virtually universal or vary by subgroup and individual. Even in the same cultural context, conventional attitudes and actions toward dogs need not be consistent: ambivalence toward dogs is recurrent (Serpell 2016).

People are affected by the norms to which they are raised and enculturated. They learn distinct and sometimes opposing values and practices toward dogs as normal. Dogs are affected, too.

Dogs, depending on their social experience with humans and other dogs in the society around them, exhibit different attitudes and habits. Some learn to be playful and seek out pats from humans while others learn to leave behind puppyish things or avoid human contact entirely. Some dogs learn complex skills, commands, and responses. Others do not. Working dogs learn their skills from one another as well as their human partners. Humans similarly learn some of their attitudes about dogs based on their social experiences with certain dogs and people and not others. Since both humans and dogs socially learn different demeanors in relation to one another depending on the culture to which they are exposed, generalizations and expectations are bound to vary from one community tradition to another. A person may learn to fear all dogs after being bitten by one as a child, or to expect dogs to be friendly if in the past they met dogs socialized to be gentle pets rather than vicious guards or suspicious strays.

Dogs in societies with forms of pet-keeping, feeding, and veterinary medicine (which are socially learned conventions and technologies present in some but not all cultural traditions) will experience different growth, health, and reproduction patterns than dogs in societies where they are free-ranging, unfed, hunting partners without veterinary care. Humans living in societies with dogs trained to aid in food provision through hunting or livestock guarding, to fend off or attack competitors, locate landmines, guide the blind, or detect cancer by smell will have different sets of available knowledge and exhibit different nutritional, health, wealth, and survival outcomes than those without such dogs or such interspecies training and work programs (Guillo and Claidière 2020). Humans who develop or enact dog-training protocols, make and use dog-oriented products, and survive thanks to trained dog interventions encode human–canine interspecies culture in themselves and their productions (see e.g. Crawford et al. 2022).

Which cultural course is taken by a community of humans and dogs has consequences for the beliefs, bodies, behaviors, and bellies of both. Both have minds and agency and are enabled and constrained by the cultural influences that converge in their being. As Kendra Coulter (2018:62) observes, "Animals are subjects, agents, creators, and co-creators of their lives and of cultures, differently able to express their agency depending on the structure and contexts in which they live and work".

Dogs' cultural abilities and proclivities are certainly different than those of humans. Our two species have evolved different but overlapping specializations. Dogs are vastly more sophisticated smellers and humans greatly outperform dogs at learning and analyzing. Dogs survive and thrive without so long and detailed an apprenticeship as humans need, yet both are deeply social. Dogs excel at reading subtle body language that humans might overlook or overthink. Human social institutions and linguistic communications are more expansive and complex than those of dogs. Yet canines are capable of learning to understand some words of the human languages to which they are exposed. They can learn to recognize and understand a limited but surprisingly extensive vocabulary and even some syntax of their humans' language (Pilley 2013). When both dogs and humans come together in communities, both can benefit from our overlapping complementarities and build cultural creations that neither species could build on its own.

Socially learned traditions, visible in behavior and artifacts, are also invented and changed by members of many animal species without any human input (Whiten, et al. 2016). Scholars of animal behavior routinely call such invented, variable, changing, group-patterned information systems "culture". Famous examples, such as termite fishing technology in some chimpanzee societies and sweet potato washing in one group of macaques, are not as complex as human legal systems or metallurgy—although cultural complexities like these are latter-day developments that did not arise until long after human cultural abilities were in place. The potential intricacy of a cultural tradition spearheaded by any species reflects its evolved intellectual adaptations and capabilities (Fitzpatrick and Andrews 2022:1104).

Magnus Enquist, Stefano Ghirlanda, and Johan Lind (2023:12) point out that while many animals including dogs use "a sophisticated interplay between genetic predispositions and [associative] learning", human ancestors evolved enhanced abilities to mentally reorganize information sequences using culture, enabling more complex and diverse models and behaviors. They (2023:7, 20–21, 77, 123) argue that associative learning can account for the social learning documented in nonhuman animals including dogs, while the cumulative and more elaborated cultures of humans are enabled by distinctive abilities to make new generalizations by mentally examining and recombining information in different sequences for imagined situations. Rather than viewing greater learning and analytical abilities as necessarily superior for adaptation or other purposes, they point out the costs of human-style intellectual and cultural abilities that may be a clue as to why they are vanishingly rare in the animal kingdom: Having a relatively limited memory and analytical capacity avoids the baffle of having too much information and too many options flood attention when making decisions. In other words, "intelligence" beyond a certain point is often an inappropriate hinderance for animals' adaptations, and therefore unlikely to evolve. The outsized memory and analytical capacities of humans can be inefficient, but these costs were evidently outweighed during "the human evolutionary transition" by the adaptive benefits of cultural generalizing, transmitting, and diversifying.

The differences between human and other animals' cultural capacities and outcomes can be productively viewed as quantitative—matters of degree—rather than qualitative—matters of kind (Boesch 2012:232). This perspective avoids the pitfall of regarding any two species as utterly distinct and incomparable entities when we are all, in fact, evolutionary cousins and in the case of dogs and their humans, foils in cultural production. "A foil", as I defined in another context, "is a person or thing that, by contrasting with another, clarifies and builds on the other's qualities" (Lohmann 2021:137). This implies similarities and compatibilities as well as complementarities between species who serve as foils for one another. Nevertheless, dogs' cultural abilities and productions do not approach those of humans in complexity. In suggesting that dogs are not just carriers of human culture, but participants with humans in making, expressing, and transmitting elements of interspecies culture, I do not mean to exaggerate dogs' capacity for culture. They do not use language among themselves or display intricate and diverse social conventions and symbolic systems akin to human cultures.

Humankind is the cultural species *par excellence.* Cultural ability, flexibility, and reliance are much more pronounced in humans than in any other species known. Each of the thousands of different cultural traditions is an elaborate and integrated collection of ideas and symbols about what is and what should be. Elements of culture are expressed and made available for others to learn as behaviors including speech and action, and the things made and altered by behaviours. Each culture includes implicit as well as conscious assumptions, scripts, and meanings. The culture of any society is a complex web of signification that cannot easily be broken down into a list of discrete parts or traits because they are mutually implicating. Many aspects like grammatical rules are automatic habits. Cultural conventions about existences, appropriateness, and values are learned by participating in a society where they are present and promoted by conformity and cooperation. The thousands of known human languages—extraordinarily complex and flexible systems of representation and communication—are also socially learned components of culture. They, and the capabilities that allow humans to develop and deploy them, enhance the power and detail with which humans create, innovate, and share cultural representations among fellow speakers.

Humans' intricate cultural abilities have enabled insights, representations, creations, and networks that would be impossible for either wolves or dogs. Nevertheless, canines do socially learn conventions from one another and from their humans. Although cultural productions with dogs often reflect human initiatives, dogs' skills and interests—including those learned from social engagement with other dogs and non-canine community members—do provide substantial impetus. Walk time and play activities, where these cultural institutions develop, are fostered and incited by dogs as much as they are by their humans, with health, social, technological, and other consequences for both. They would be senseless and unachievable without members of both species contributing to their realization.

Without dogs trained to take on assigned roles (like pet and livestock guard), use provided artifacts (like harnesses and doghouses), and evolve preferred characteristics (like curly tails and piebald coats), in response to artificial selection, these cultural manifestations would not be possible. It is inaccurate to treat such cultural expressions in dogs, or those in humans or other ecological elements that are realized through canines, as purely human cultural interventions or creations because they come into being through the bodies, social learning, and agency of canines as well as people.

Dog breeding and breed standards reflect particular social groups' cultural abilities and preferences. For example, Mongolian herders favor black dogs to easily distinguish them from wolves (Bamana 2014:3). Italian shepherd humans, in contrast, strongly prefer white shepherd dogs. Although they cannot control breeding, their cultural preference is expressed in canine phenotypes and genotypes through eugenics-style culling (Coppinger and Coppinger 2016:179–186, 213). Generations of selecting and favoring the reproduction of individuals with this coat color, based on a cultural preference, effects more than social learning of symbolic systems. This "phylogenetic enculturation" has in effect written culture in the DNA of dogs (Hare et al. 2002:1636).

Oka pastoralists in Siberia rely on their older dogs to model desired behaviors toward game versus livestock to puppies. These humans kill young dogs who show little aptitude for their assigned tasks or inappropriate aggression.

"Culling dogs in their first year serves as a form of selective breeding" (Oehler 2021:768). Since dogs learn local expectations from the modeling of what others do and are rewarded for in their society, culling those who do not conform is also a form of selective enculturation, for it removes their cultural influence in their canine–human society. The effects are not restricted to the canines. For example, with historical shifts in culture, street dogs in Ottoman Egypt were offered and denied divergent ecological, economic, and social roles, with not just attitudinal, but also life and death consequences for members of both species (Mikhail 2014:67–106). Indeed, cultural changes in the history of Greyhound-keeping in Britain resulted in both genetic and behavioral changes in these dogs, with influences circling back to their human "masters" (Russell 2018).

In the shorter term, growth, nutrition, and longevity outcomes are influenced by the kinds of treatment dogs receive. Through culturally variable provision or control of diverse sorts of shelter, food, and medical assistance, dogs' bodies encode the distinctive human–canine interspecies culture that such conditions manifest. In this way, aspects of anatomy and physiology can be modified in accord with culturally distinctive conventions characteristic of different societies. The bodily condition of individual dogs, like artifacts, physically manifest behavioral and ultimately mental culture (Lohmann 2010:219). Even human ideas about dogs—such as the notion in the Philippines and elsewhere that they sense and warn of evil spirits—would not arise without canine participation in our societies (Stanyukovich 2023:418).

The category of human–canine interspecies culture calls to attention cultural flows, manifestations, and consequences not just in human and dog memories, traditions, societies, and behaviors, but across whole ecosystems. As Pat Shipman (2015:228) argues, "the technical advance that made humans so irresistible and so invasive…was in part their ability to form this unprecedented alliance with [dogs]". Paleolithic human–dog societies conferred multiple advantages to both species through the cultural expressions in which both species contributed necessary parts, such as human-loaded and dog-carried firewood and food (Germonpré et al. 2023). Myrdene Anderson (1986:7) observes that dogs are necessary for the reindeer herding way of life of the Saami of Lapland. Anderson calls this "dog culture … manifest in the herder/herding dog/reindeer triad and in the interpenetration of assumptions concerning child/dog enculturation" (2016:32). As John Marzluff and Tony Angell (2005:69) point out:

> when humans interact with other social species, who themselves have the ability to evolve culture, then simple feedbacks from a culturally evolving 'environment' can stimulate rapid cultural evolution in humans. We term the reciprocal adjustments in two or more species' cultures 'cultural coevolution'.

Given the antiquity, ubiquity, and homey familiarity of dogs, focusing on human–canine interspecies culture offers an easily accessible peek around the anthropocentric blinders that so easily constrain our vision and understanding. It starts by looking at societies and cultures as comprising not just humans, but other social species like dogs when they are present. One fine ethnographic example of this perspective is Yasmine Musharbash's (2017) description of Warlpiri people and dogs in the Aboriginal Australian community of Yuendumu, Northern Territory. Warlpiri name their dogs, but unlike human names, dog names are never used to call a dog, and dogs' names do not become taboo following the dog's death. They scarcely direct speech at dogs, and do not teach dogs to obey commands. They do not pet, snuggle, or groom their dogs. Toddlers are encouraged to treat puppies roughly, including throwing, hitting, dragging by one leg, and kicking. They do this to enculturate dogs to not attack people, and people to have confident authority over dogs.

In the past, Warlpiri and other Australian Aboriginal peoples took puppies from the dens of dingoes and nursed them from their own breasts. Canines' propensity to alert the community to the arrival of strangers and supernatural monsters was prized. Warlpiri people also find the antics of the dogs in their camp entertaining. They describe dogs as companions and family and mourn their deaths. From this sample, we can see that cultural notions about dogs (in people) find expression in habitual behaviors and distinctive artifacts. However, these cultural forms are not strictly human since they arise from the presence and participation of canines in society with Warlpiri people. They also enter dogs' bodies in the form of conventionally allowed human milk, trauma from normative mistreatment as puppies, and in learned expectations of traditionally minimal tactile and verbal interactions and responses as adults.

When Aboriginal Australians took dingo pups from dens, they shaped their experience and learning during a key formative period. These pups remained with their humans during their youth but then departed to find mates and raise their young in the bush. Considering these facts, Adam Brumm (2021:24) suggests that:

> dingoes reared by humans behaved differently to purely wild dingoes: they were not subservient and dependent like domesticated European dogs, which are scarcely able to survive without human assistance, but they also did not behave like dingoes that had been socialised by wild dingo parents and trained how to hunt from an early age. Life with humans, I contend, modified and reshaped the brains and behaviour of these pups.

Dogs clearly learn manners and skills from one another and from humans (Coren 2004:227). Recognizing that this is so challenges assumptions that dogs all have similar tendencies and capabilities, and that any differences are based on genetic variation and supposed breed characteristics alone. It also warns against facile breed stereotypes that amount to racism toward dogs.

Sorting biological from cultural capacities and variations in canine attitudes and actions promises to be a valuable research avenue. Simply by expecting possible cultural differences in dogs, as anthropologists expect to a greater degree in humans, we can improve the sophistication of canine studies in multiple fields.

Research comparatively assessing species-wide chimpanzee, dog, and human abilities to communicate intention and share common attention, through such means as pointing, suggests that while chimps only learn these techniques with difficulty, humans and dogs share an inborn facility in this area (Hare and Tomasello 2005). However, just as humans in different cultures do not use the same gestures or body parts to point, but rather learn these conventions by enculturation, dogs too learn attention to humans and to particular communicative conventions by participating in specific human–canine societies (Udell, et al. 2010). In mountainous inner Asia, for example, Alex Oehler (2021:769) documents hunting dogs deciding which scent trails they are to pursue by following their human partners' line of sight. Clive Wynne (2019:40, 118) stresses that following pointing gestures is a socially learned ability that arises in multiple animals, including wolves, when they grow up in dependent social interaction with humans who use those gestures. Wynne (2019:125–126, 171) identifies another quality as key to understanding the human–canine bond: dogs' unusual capacity for emotional engagement and affectionate interaction that can cross to other species if appropriate social exposure to them occurs:

> Although having the right genes is crucial to dogs' loving nature, it is far from the whole story of who they are. Individual dogs grow up to be a range of things: the loving partners that share our homes, dangerous and aggressive beasts that people have no desire to come near, and even wild animals that are afraid of people and never seek their company. Life experience is the clincher
>
> (Wynne 2019:186).

In other words, dogs, like people, differ depending on the enculturation they have received. And enculturation is an ongoing process. Ádám Miklósi (2015) warns that dogs respond to subtle cues from their human handlers, which can be misinterpreted to indicate greater capabilities than they actually have. This phenomenon, known as the Clever Hans effect, also indicates that dogs and many other animals continually extract and use more and different information from social engagement (even across species) than researchers might expect. Studies of dogs' abilities and habits need to take these differences into account to avoid the kinds of cultural bias that plague psychological studies that make universal pronouncements about human characteristics without taking cultural inputs and variation into account (see Lende and Downey 2012:404).

Some studies have sought to identify cross-cultural differences in dogs. In a comparison of German Shepherd dogs in Hungary and the U.S.A., owners' assessments of their dogs' abilities and inclinations were found to be more or less the same (Wan et al. 2009). In another, Border Collies and Labrador Retrievers in Austria, Hungary, and the U.S.A. were given identical cognitive tests and found to be largely similar (Szabó et al. 2017). However, the cognitive abilities tested are perhaps not socially learned skills and inclinations. Furthermore, these three countries are relatively similar culturally. I expect testing how easily Āpia street dogs can learn the command "come" in either English or Samoan, or measuring how at ease they are with humans, might well produce different results from identical behavioral experiments done on Arapesh village dogs and trained Japanese performing dogs of similar genetic makeup. Likewise, dogs trained to detect illicit substances for the Canada Border Services Agency would be expected to be more effective at that task than dogs of the same breed but different training. Training is culture: it manifests social learned information from a distinct and variable tradition and its consequences. Dog training is human–canine interspecies culture and central to our joint ventures. Cultural abilities and preferences are inscribed in both active and passive training.

Dogs' mental models, behaviors, and even bodies are altered through linguistic and non-linguistic communication of socially learned information and its effects between culturally distinctive human and canine associates (Kulick 2017). Different skills, behaviors, and attitudes in dogs are demonstrably dependent on different training methods—cultural variation that is visible in both the archaeological and ethnographic records (Burtt and Bethke 2020). Punishment-based dog training is associated with a higher degree of behavioral problems than reward-based training (Hiby et al. 2004). Even dogs' degree of optimism—inferred from how quickly they investigated bowls that they knew might or might not contain food—has been positively correlated with reward-based training—in other words, the content of their enculturation (Casey et al. 2021).

Placing Interspecies Culture in Theory

I use the term interspecies culture to refer to socially learned information that manifests as, or results in mental representations, behaviors, physiology, anatomy, or artifacts, in which multiple species are involved. Culture crosses species lines in cases such as training, breeding, and tools of interaction. It turns up in "interspecies communication traditions", like African honeyguide birds who respond better to local than to foreign human calls (Spottiswoode and Wood 2023:1155). It turns up in novel varieties and decreased genetic diversity in agricultural crops (Louwaars 2018). It can be recognized in elements of a living thing's genotype that was shaped by a culturally influenced interaction of more than one species, such as when a culture-bearing species domesticates, selectively breeds, or genetically modifies another (Altman and Mesoudi 2019). It turns up in technologies that would not exist without another species in mind, like reindeer lassos (Ingold 1993).

My training in general anthropology and ecological awareness promote attention to humans and culture as components made up of and enmeshed among other interdependent parts of nature. With this background I independently developed the concept of interspecies culture to theorize the involvement of both humans and dogs in forming and expressing divergent traditions of socially learned information.

I began by thinking of differences in terms of different demeanors that dogs learn depending on the ways that people habitually treat them in different cultures. This progressed to realizing that these culturally specific treatments have health and developmental impacts. I then realized cultural engagements would have genetic evolutionary consequences by shaping which traits were selected for. I noted that since dogs are participants in these developments they cannot be attributed solely to humans. I reasoned that humans would not develop many of their techniques and technologies without dogs being there as intended targets or cooperators. Finally, it occurred to me that not just dogs, but their humans as well undergo changes in their mental, behavioral, health, developmental, genetic, and artifactual expressions of culture that would not exist without both species participating.

Humans and dogs are not alone in ecological communities, and cultural expressions involving more than one species extend beyond these two species. Indeed, neither humans nor any other species, culture-bearing or otherwise, acts or even exists alone, but necessarily in webs of ecological interdependencies with other living things within and outside of our bodies. Without the interactions of the plants, animals, fungi, and microbes that sustain humans, for example, there would be no anthropogenic cultural creations or diversity. Seen in this light, all culture is interspecies culture. That said, as Laurence Carucci (this volume, Chapter 9 note 5) points out, not all participants in cultural productions are equally active or intentional as creators of culture. To recognize the interaction of multiple species as a condition of all cultural productions is not to attribute the same creative, thinking, learning, and teaching roles to all the species whose interaction generates and changes culture.

As an anthropologist in the holistic tradition that integrates linguistics, biology, archaeology, and ethnography to understand humankind and human variation, I am accustomed to making culture central, seeing it in the long term and in multiple forms including artifacts, utterances, and body states. My ecological perspective shows all living things in the biosphere as existing in interdependent and changing relationships with one another and with nonliving things. From this vantage point, humans and our cultures are obviously parts of the global ecosystem, such that considering culture or humans as somehow external, or opposite, or above nature is inaccurate. My experiences with dogs during ethnographic fieldwork, as described in Chapter 8, made me aware that dogs act and develop differently depending on the culture of the human–dog community that they live in. My pet rescue dogs made it obvious to me that dogs not only learn different demeanors and habits, but express different health and weight outcomes depending on how the people around them routinely treat them.

I came up with the term interspecies culture while writing the first draft of what would become this introduction to refer to the mutual effects I was observing in dogs and their humans. I presented it at a meeting of the Canadian Anthropology Society (Lohmann 2022). As this volume project developed and went through peer review on the way to acceptance for publication, I became aware of several other publications that made use of the phrase in narrower but compatible ways with mine.

In the first, Dan Parker, Kylie Soanes, and Stanislav Roudavski (2022) use the phrase "interspecies cultures" in discussing urban design efforts that promote the abilities, wellbeing, and expression of multiple resident species' distinct local and current cultural traditions:

> We see this as an opportunity to cultivate interspecies cultures which curate non-anthropocentric interactions and foster beneficial relationships between humans and non-humans. This is possible because both human and nonhuman animals continually reconstruct their cultures. Cultures develop across generations, emerging as beliefs, knowledge, skills, traditions, or practices .... As an example, where humans taught birds novel foraging techniques demonstrates, such cultures are influenceable and can spread rapidly throughout populations
> (Parker, Soanes, and Roudavski 2022:188–189).

Parker, Soanes, and Roudavski's notion of interspecies cultures, like mine, draws attention to cultural expressions that involve more than one species. Their focus on developing built environments that satisfy cultural and biological requisites of interacting species represents a practical extension of the concept. Their conceptualization of culture, also like mine, includes socially learned mental representations and their behavioral and artifactual expressions. They acknowledge that interspecies cultural expressions have ecological and biological consequences in habitats and individual survival, which is also congenial to my formulation. However, they do not explicitly define what they mean by "interspecies cultures".

Cédric Sueur and Michael A. Huffman (2024:821) presented the "concept of co-culture as a novel framework for understanding the mutual cultural evolution between two animal species, including, but not only, humans". They use the phrase "interspecies culture" in their title as an explanatory synonym for co-culture. They (2024:822) define co-culture as "mutual cultural evolution and interaction between different species, like humans and animals, leading to collective cultural development through shared adaptations and behaviours resulting from direct or indirect interactions within the same environment". They point out that these two-way cultural influences may be symbiotic and can affect both behaviors and, in the long term, genetic traits. Sueur (2022:96) pre-cites this work, noting that interspecies communication and mixed species groups in proximity are key drivers of co-cultures.

Co-culture seems compatible with my usage of interspecies culture, although our emphases and extensions vary. They use culture to mean socially learned behaviors, especially those relevant to ecological adaptation and niche construction. They do not directly address mental models or artifacts as forms of culture. The "co-ness" in co-culture refers to interspecies learning of cooperative techniques. The sense of interspecies culture that I use is somewhat broader: I refer to information in any of its manifestations or effects including mental representations, behaviors, artifacts, phenotypes, ranges, and genetic characteristics, that was once socially learned and that requires the participation of more than one species to exist.

A third recent article uses the phrase "interspecies cultural practice" to refer to situations in which more than one "species come together to co-produce a shared space of interaction that relies on both of their ongoing participation to endure" (van Dooren et al. 2025:17). Thom van Dooren and co-authors deploy this usage to highlight the fact that socially learned ways of life are known among a wide range of animals, that some of these include behaviors learned from interactions with other species, and that these can be disrupted and lost when any connected species goes extinct or ceases to participate. Their usage is compatible with but much narrower than my category of interspecies culture.

Cultural phenomena can be co-productions of more than one species. The ways in which this occurs constitute a productive zone for research. This reflects in part simply taking seriously long-established scientific precepts about the entwinedness of humans and culture within the rest of nature. The shift in attention that this brings, makes obvious how anthropocentrism limits our understanding of ourselves and the world. Perhaps ironically, by focusing on our own kind, humans cut ourselves off from understanding our broader nature. Anthropology's comparative study of multiple cultures in different places and times unmasked the dangers that ethnocentrism and presentism pose to our understanding of humankind as a whole. Similarly, ecological cognizance of the mutual instantiation of multiple living and nonliving things through their systemic and changing interactions lays bare the conceptual and existential dangers of human exceptionalism framed by ecological ignorance and denial.

The interspecies culture concept is in keeping with multispecies approaches in anthropology and other social sciences and humanities. Their purview extends the study of humans and ethnic groups to include at least some of the other species that are present and vital in their communities:

> Multispecies ethnography [for example] …. acknowledges the interconnectedness and inseparability of humans and other life forms, and thus seeks to extend ethnography beyond the solely human realm.

> Multispecies investigations of social and cultural phenomena are attentive to the agency of other-than-human species …. to overcome anthropocentrism … and to highlight the intersections between ecological relations … and cultural representations
>
> (Locke and Muenster 2015).

I use the term interspecies rather than multispecies culture because it emphasizes the interactive relationship between different kinds of living things rather than simply their co-presence as vital to cultural instantiation. I came upon it as an anthropological extension of the ecological truism that humans, like other living things, exist through interchanges with other living and nonliving substances and things. Our cultures are no exception. Therefore, we can be misled by studying any species in isolation. This scientific recognition long predates so-called posthumanism, the animal turn, and other "turns" of academic fashion and the multispecies ethnography that such movements fortunately helped inspire (e.g. Chao 2022).

Those who use anthropological methods to understand the interactions and consequences of multiple species in common societies and ecosystems help us to recognize "the human as a kind of corporeality that comes into being relative to multispecies assemblages" (Ogden et al. 2013:6). The same can be said for "the dog" and for the cultures that include dogs and their enactments. Including members of multiple species as subjects in ethnographic and archaeological descriptions can help anthropology move beyond anthropocentrism, to a recognition that societies include more than the single species of *Homo sapiens* (Kirksey and Helmreich 2010). This is all the more important for a science-based understanding when the ethnic cultural points of view toward animals so documented are themselves anthropocentric (Morris 1998:6, 85–89). As John Hartigan (2021:856) observes, multispecies ethnography can document ecological effects of multispecies associations that members of none of the species involved are aware of. And, as Martin Drenthen (2023:292–293) points out, interspecies communications and the social learning they convey can generate stable traditional understandings for co-existence between humans and dogs' wild cousins:

> If farmers consistently protect their sheep so that it becomes so cumbersome and risky for a wolf to take sheep that it becomes unappealing, wolves will learn to adapt their behaviour, provided of course that there is enough wild prey to hunt. Experience … shows that, after a few generations, young wolves will grow into the culture of the adult wolves, in which kept livestock are ignored …. It might be tempting to think of wolf fences as hard borders between culture and nature, but that would bring us back to the outdated dualist mode of thinking. Rather, these fences should be thought of as means of communication between species that help establish low-conflict human–wolf coexistence.

Humans are parts of canids' environments and vice versa. Our species are interacting, culture-using engineers of one another's survival, health, behaviors, and understandings. Interspecies culture, as I have formulated it, directs our attention to ways and cases in which socially learned information in all its forms and effects necessarily involves more than one species to be created and maintained. I have found it enlightening for describing the diversity of dog–human productions. It may be found to be a useful concept for recognizing and analyzing other cross-species productions that involve culture.

Examples of Human–Canine Interspecies Culture in this Volume

Edited volumes such as this do not speak with one voice or present a single perspective. Each of this book's contributors brings his or her own approaches and insights to the anthropological study of dogs. Most have found my formulation of interspecies culture useful or effective, while some prefer other analytical conceptualizations. The empirical and theoretical diversity that these contributors bring to this project is a strength. In this section I summarize the coming chapters, emphasizing elements that, in my estimation, exemplify human–canine interspecies culture. I also indicate how each author has engaged with the book's titular notion in their own analysis.

In Chapter Two, "What Arapesh Dogs Mean: The Figure of the Talking Dog on New Guinea's Sepik Coast", Lise M. Dobrin reports that dogs feature in local stories as heroes able to bridge gaps between different language groups. They are also valued exchange partners who give their humans the meat they catch in the hunt, and in return receive care and feeding. Dogs are reputed to have formerly been able to talk, and indeed, to have been able to change into human beings. Dobrin opens her chapter comparing the different health and behavioral outcomes of an Arapesh puppy she raised compared to a littermate raised in the village with other local dogs. She identifies this as exemplifying different interspecies cultures. Her focus, however, is on how Arapesh people represent dogs, with particularly lovely detail on the language.

Dobrin's analysis shows that dog–human interaction in the Arapesh cultural sphere provides people with tools for thinking through the possibilities of social connection across difference. An elderly woman told Dobrin that her parents instructed her not to hit dogs, because they are really people, and their cries reflect linguistic utterances. Yet people also frequently express disappointment in their dogs, who do not provide or share as freely as their ideal reputation would have it. This only makes them resemble real people all the more in Arapesh estimation. Here, the mutualistic interspecies relationship of the canine–human hunting partnership helps people conceptualize and facilitate exchange relationships and sociality between humans who speak different languages. In turn, dogs respond to the treatment that results from their humans' admiration, achieving better health and security than they could have on their own or in different cultural settings.

In Chapter Three, "Dogs in Majuro: Pests, Pets, and Perils", Jordan Prokosch describes an urban dog situation, but one without animal shelters, customary training, veterinary practices, or companionate pet keeping. In the Marshall Islands' capital city of Majuro, Prokosch found most dogs to be thin and diseased strays surviving on garbage. They are unprotected from the cruelties of children and are slaughtered for the clandestine cooking fires of young men. They are unfriendly to people in return but generally too weak to be a threat. Other dogs, classified locally as pets, are unsupervised household guard dogs who, while cared for and healthy, live primarily outdoors where they not only protect the wealthy from the poor, but also threaten and frequently bite passersby. Prokosch, citing my own approach, takes from the Majuro case that while dogs are not possessed of their own cultures in the complex sense that humans are, they do respond strongly to the cultures of the humans around them.

Prokosch describes with evocative clarity the Marshallese attitudes that surround humans' forms of living near and with dogs. Noting that concepts of loving, responsible care and empathy are central to identity and practice in this, as in many other Oceanic cultures, he observes that such directives do not traverse species lines. Prokosch intriguingly suggests that people's dismissal of obvious suffering of dogs and other animals while defining themselves and their culture as empathetic is a form of denial. Seriously pretending that non-human animals do not really suffer allows people to neglect and even take pleasure in harming them without having to endure feelings of repulsion, regret, and loathing of oneself, one's fellows, or one's culture. Dogs are some of the most numerous and sentient victims of this foible. Obviously, such denial, inconsistency, and avoidable cruelty is present in many cultures, but takes different forms.

In Chapter Four, "More Aloha for the Dogs: The Cultural Context of Dogs in Hawaii", Lynn Morrison uses my concept of "human–canine interspecies cultures" to refer to different and malleable interactional practices and their consequences on opposite sides of the big island of Hawai'i. In the east, many dogs participate in pig hunts. Their humans routinely leave them chained in yards where they are starved in the belief it will make them more enthusiastic hunters. Abused and abandoned dogs are collected and brought to a local shelter. Shelter workers attracted by their love of animals are daily obliged to alternately care for and kill dogs not collected or adopted within 72 hours. There is a continual flow of dogs through these institutions because of local throw-away attitudes among these humans toward dogs. In the west, a greater proportion of dogs are pets rather than employed (or enslaved) to hunt and guard for their humans. There is less neglect and abuse, and spaying and neutering are more widely practiced. Fewer dogs are admitted to the shelter. Shelter dogs are more likely to be retrieved or adopted and have better average health. Although the same policies apply at this shelter, shelter workers are not expected to kill their excess canine charges so frequently.

Sampling the stress hormone cortisol in both shelter dogs and shelter workers, and taking blood pressure while interviewing the humans, reveal the different degrees of harm inflicted on both species' lives and well-being depending on customary attitudes and practices governing human–dog interactions. Both shelter workers and dogs showed significantly fewer indications of chronic and situational stress at the Kona shelter in the west than at the Hilo shelter in the east.

The sad condition of the dogs in the east—emaciated, injured, and fearful—and of the shelter workers trapped in having to routinely kill the very dogs they care for day in and day out—is moving to read. So too are the testimonies from beleaguered shelter workers quoted from interviews, who on top of everything else, face blame for euthanasia policies that are driven not by them, but by community standards. Morrison illustrates profoundly that the bodies, behaviors, and mental health of both humans and dogs are divergently affected by the different streams of human–canine interspecies culture within broader Hawaiian society today. She recommends culturally tailored interventions, starting with positive and loving discourses drawing on the deeply meaningful Hawaiian values of active and compassionate caregiving to others and the land—known in Hawaiian as *aloha*—to better the situation for these dogs and their humans.

In Chapter Five, "Is a Dog an Artifact? Archaeological Investigations of Dogs and Their Humans in Oceania", Karen Greig introduces the wide sweep of dogs across Pacific times and places. Her title alludes to a tendency in some cultural settings to view dogs as their humans' utilitarian possessions rather than as sentient and sovereign beings. As Greig shows, without the complementarity of other perspectives, this understanding of dogs is unnecessarily impoverished. Dog lifeways—not just human ones reflected in dogs—can be discerned from archaeological as well as ethnographic evidence. Taking insight from multiple cultures' models of reality can help archaeological interpretation of how and why past peoples thought and acted as they did toward dogs, their corpses, and products such as dogs' tooth jewelry and dog-skin cloaks. Merely pragmatic explanations may be trite or misguided by foreign and anachronistic assumptions.

Greig points out that while archaeological dog finds have typically been regarded as indexes of the behavior and culture of humans, they also provide information on human–canine interspecies culture. She deploys the concept as part of recognizing physical as well as mental and behavioral manifestations of culture that come into existence by crossing species lines. Dogs are not artifacts in a merely anthropocentric, utilitarian sense because they have jointly participated in multiple past and present societies. From their humans' points of view, they have not participated on equal terms; however, dogs and humans both exhibit sentience and agency in realizing joint cultural achievements. We influence one another in mutually causal ways. Cultural differences in past and present societies that include both species express a range of relationships between the two.

In Chapter Six, "Dog or Dog-Gone? A View from Four (Almost) Dogless Islands", Richard Feinberg and Cathleen Pyrek describe dog–human conditions on four small Polynesian islands: Anuta and Taumako in the Solomons Islands, Nukumanu Atoll in Papua New Guinea, and Atafu Atoll in Tokelau. They show that it is often problematic in the long term for dogs and people to share small islands because real or perceived competition for scarce resources tends to eclipse the benefits people attribute to having dogs among them. Feinberg and Pyrek productively engage with the "human–canine interspecies culture" concept by pointing out that "culture" has been used in many ways and the purpose of definitions is usefulness for achieving an objective. Those accustomed to defining culture as a symbolic system that excludes social systems may find more inclusive definitions uncongenial because they do not suit the purposes for which their familiar definition was designed. Feinberg and Pyrek find interspecies culture as I have defined it useful for calling analytical attention to distinctive beliefs, actions, and physical effects of socially learned information manifesting through interactions across species. As an example, they mention Samoan dogs' abilities to avoid being hit by cars, but many other illustrations can be found in their chapter.

Tiny Anuta was dogless during Feinberg's multiple periods of ethnographic research. The people regarded dogs as competitors for food with little use. One dog reputedly brought to the island in the past had attacked chickens and threatened people, and this solidified Anutan attitudes against having dogs. Anutans have encountered dogs kept for security in the city of Honiara on Guadalcanal. These dogs were not ordinarily fed much or treated with kindness. They relied on scavenged food on the streets, where they were fearful of, or aggressive towards humans, who kept them at bay with threats of thrown stones. Most Anutans who met dogs that had grown up under such circumstances were understandably afraid of dogs and saw little point of bringing them home to their rural island.

On Nukumanu atoll, dogs were also absent. Nukumanu people who encountered dogs in urban centers elsewhere in Papua New Guinea, such as Port Moresby, had similar experiences with, and attitudes toward dogs as Anutans. Taumako, about five kilometers long and three kilometers wide, did have resident dogs. Some of these lived with people, although not within their one- or two-room leaf houses. Taumako dogs foraged to supplement the leftovers that their people fed them. Some dogs wandered off to become feral in the island's hilly forests, where they attacked people's pigs, and were in turn hunted down and killed to reduce this threat. Dogs also featured in Taumako spirit beliefs, where a dog spirit is the protective totem of one of the descent groups. On Atafu Atoll, Feinberg and Pyrek found no dogs at all. Atafu emigrants to nearby Samoa encountered stray dogs on the streets of Āpia, where they sometimes threatened humans, who responded by throwing rocks. Feinberg and Pyrek's Atafu hosts in Āpia kept a watchdog, whom they fed table scraps and treated as a working dog rather than a pet. In response, this dog did not let people near: not even her owners. However, when she had puppies, the household head bonded with one.

In these examples we see again that there is a cultural exchange going on between dogs and humans. People in Honiara typically learned to assume that dogs are vicious, hence ordinarily used them for security and fended them off with violence rather than taking them in as pets. Local dogs develop expectations and behavioral scripts according to their training and other experiences. When Anutans encountered this foreign human–canine interspecies culture, they came to assume that viciousness is typical of dogs, making them less likely to bring dogs back to Anuta. With other cultural tools, Anutans could train dogs not to kill chickens, not to threaten people, and indeed to join the cats in killing rats. Dogs participating in such a system would also play an active role in training up the expectations and responses of their humans.

In Chapter Seven, "Dogs without Their Humans: Occurrences of 'Wild Dogs' in Oceania Prior to European Colonization", Loukas Koungoulos explores the evidence of dogs who have colonized wild areas and declared independence from humans before the arrival of novel breeds. He uses the term "wild" to designate such populations and differentiate them from "feral" dogs, defined as remaining dependent or on the margins of human societies. Only on two of the largest landmasses were there stable populations of dogs living without any regular interactions with humans: the Australian Dingo and the New Guinea Highland Wild Dog. Nevertheless, Koungoulos quotes multiple historical accounts of wild dogs on other western islands of Melanesia, referred to as Near Oceania because it is closer to the Asiatic first source of in-migration for both humans and canids. Wild dogs are also attested on larger islands further east and south—a later-settled region called Remote Oceania. However, in none of these places did canine societies clearly exist as sustainable populations without immigrants from domestic human settings.

Koungoulos suggests that the paucity of wild resources on smaller islands likely prevent such self-sustaining canine groups. He does not make use of the human–canine interspecies culture concept in describing these situations, for the dogs he focuses on are no longer in any regular contact with humans. However, feral dogs and truly wild dogs, although they or their ancestors have left or been abandoned by their erstwhile people, still manifest human–canine interspecies culture in their very physical presence in places where they could not have reached without cultural engagements with humans.

In Chapter Eight, "Cultural Differences in Dogs in Central New Guinea", I tell how I was struck by differences in the ways the people who hosted my ethnographic research, the Asabano, treated their dogs in comparison to neighboring Telefolmin and Omai communities. I noticed that Asabano dogs were often struck with a stick, raising a pitiful yelp when they came too close to food, or made any noise in the house. Many dogs were thin, and some were mangy. I saw a puppy treated inconsistently, being kissed and thrown harshly in close succession. Asabano dogs skulked and were skittish towards humans. In contrast, Telefolmin and Omai dogs were plumper and approached humans more confidently.

All three of these peoples valued their dogs for hunting, and Asabano dogs who won a reputation as helpful hunters received better care and appreciation than others. But cultural differences in the interactions between the species in each community went beyond there being different words for "dog" in the three languages. Dogs treated people differently and people treated dogs differently depending on the culture that governed interactions between the species. Canine and human nutrition and health were both influenced by the forms of interaction and cooperation found in each community. Local marsupial populations also felt the consequences of human–canine cooperative hunting. Witnessing these variations that depended on the human cultural group dogs were affiliated with alerted me to the existence of interspecies culture and its physical as well as its behavioral reach.

In Chapter Nine, "Going to the Dogs with Enewetak/Ujelang Marshall Islanders", Laurence Marshall Carucci describes how Enewetak islanders had been relocated to Ujelang so that the U.S. American government could test nuclear weapons on their tiny and verdant homeland. No dogs were kept on the inhabited islets of Ujelang Atoll when Carucci first arrived for ethnographic fieldwork. The atoll's former dogs had been exterminated and eaten to better enable the resident cats to bring the newly introduced Norwegian rats under control. A population of dogs was, however, introduced to an uninhabited islet where they scavenged for seafood at low tide. Young men sometimes killed and roasted these dogs to eat at drinking parties, but higher-ranking people never ate dog. After returning to their irradiated home atoll of Enewetak, people kept dogs. These dogs led difficult lives because the community was now dependent on imported food supplied by the United States, and this allowance did not include sustenance for non-human animals in the community. Dogs lived on scraps and made an understandable nuisance of themselves.

Carucci had first encountered dogs on the urbanized government atoll of Majuro, where he, like Prokosch (this volume), was horrified at their treatment. Humans shooed dogs by throwing coral rocks at them amidst hisses, often causing them injuries. This behaviour was also prevalent among the Enewetak people. Furthermore, children carried and swung puppies around by a single leg and "played drowning games" with them. In both urban and rural communities in the Marshall Islands, Carucci saw similar treatment of dogs, with the exception of eating dogs, which was restricted to his field site. In this complex example, not just human bodies, but also canine bodies were irradiated and starved by humans enacting

Cold War American culture. Dogs living in marginalized conditions with marginalized people suffered the effects in their level of nutrition and care, abuse, and genocidal annihilation in response to cultural introductions of other species and distinct cultural attitudes toward them. The food insecurity of the people, and cultural allowances for mistreating and eating dogs, had definite results for both human and canine bodies and behaviors. I find such phenomena illustrative of what I am calling interspecies culture and expressions of culture in dogs. However, Carucci finds such usages an untoward extension of the culture concept as it has traditionally been used in cultural anthropology because of differences in how humans use arbitrary symbolic systems compared to other living things like dogs.

In Chapter Ten, "A Samoan Odyssey: Mysterious Spirit Dogs and Interspecies Enculturation", Micah Van der Ryn reports an unusual experience of being followed by a pair of particularly healthy-looking dogs on a multi-day walk around Savai'i Island. He guessed, but was not sure, that they belonged to a family he had stayed with the night before they joined him walking in the pre-dawn darkness. Van der Ryn suggests that the dogs may have been motivated to accompany him on his journey out of empathy with his host's concern that walking alone in the dark was dangerous. One local told him that they were not really dogs at all, but helping spirits, an idea that Van der Ryn demonstrates as congruent with Samoan spiritualism. Drawing on the notion of human–canine interspecies culture, he endeavors to explain these dogs' unusual behavior in terms of the values and beliefs that both humans and family dogs are enculturated with in Samoa, including "loyalty, service, compassion, and respect for social hierarchy".

As socially appropriate behaviors are rewarded, some of a society's cultural information finds its way across species lines to dogs. Van der Ryn notes that the stray dogs of towns like Āpia are generally poorly treated because they lack identity as a member of a human family but nevertheless live near them and rely on them for food. He contrasts such stray urban dogs with those owned by people, who are in better health because they exist as canine members of a human family. Unlike other animals, family dogs are named and remembered when they die. Samoans generally abhor eating dogs, and in this they contrast themselves with Tongans.

Also telling is Van der Ryn's discussion of how traditional and more recently introduced architectural styles affect both humans and their dogs. Samoan houses of the time-honored style are more open than new-style houses enclosed in solid walls and fenced yards. These styles reflect and engender different cultural attitudes that affect the behavior of both dogs and humans. For families who move into sealed-off dwellings, Van der Ryn indicates that the dogs absorb the changed attitude of the people, becoming more territorial and taking on more aggressive styles of guarding the space against outsiders. This contrasts with the increased familiarity among inhabitants that one sees in Samoan villages that are both physically and socially more open spaces. Again, in this Samoan example we can see how dogs' and humans' behaviors, bodily health, nutrition, development, dispositions, and ideas about reality are shaped by cultural notions that arise in interactions between the two species.

Anthropology's study of humans is bound to teach us a lot about dogs because they are so often around people. This volume presents many examples showing that dogs, far from being identical in form, behaviors, skills, and attitudes regardless of the culture in which they live, vary cross-culturally. In Chapter Eleven, "Sniffing toward the Future: Accessing Interspecies Enculturated Dogs' Perspectives", I identify a potential future direction for the anthropology of dogs suggested by this recognition: turning the discipline's methods for studying cultural difference more directly to dogs themselves.

Dogs learn and show distinctive perspectives on reality, characteristic of the societies in which they live. Therefore, understanding any dog's behavior, abilities, habits, and motives requires taking their cultural contexts and learning into account. Efforts to understand dogs' mental lives, propensities, and abilities through their distinctive biology and sensorium establish a baseline shared across dogkind. We can begin to relate to canine perspectives by recognizing that they are supported by four legs rather than two and a smell-centric rather than a vision-centric experience of their environment. However, missing here is theory and method for recognizing and documenting culture's effects and differences. These, for example, make Majuro feral dogs' worldview and characteristic actions different from those of wild dogs in New Guinea who live independently of humans, or of Samoan family dogs who learn and enact some of the cultural values of their people. Thanks to a century and a half of anthropological research, we know that human ideas, behaviors, and artifacts differ more based on what groups socially learn than on their racial or genetic differences; just as racial profiling does not provide accurate information about how any humans will act or think, species and breed stereotypes can't explain why some dogs cringe, others approach with wagging tails, and still others perform intricate moves when asked to do so in a particular language.

Stereotypes based on biological similarities across the human species obscure the very significant differences in perspective between different ethnic groups. To understand these, anthropologists use long-term exposure, interaction, communication, and participant-observation. Anthropological methods like these are designed to learn about other groups' culture-based understandings of the world through close and empathetic engagement with their societies. While the separation in biology, mentality, and understandings between humans and dogs is far greater than the separation between humans of different cultures, our two species' abilities to live and work together suggest that they are not insurmountable. Although we cannot exactly interview dogs with language, I suggest some ways of adapting ethnographic and other anthropological methods to better understand their culturally distinctive worlds of experience and convey these across species lines.

I guided the development of this volume over three annual sessions at Association for Social Anthropology in Oceania meetings, concluding in Kona, Hawaii in 2023. I wanted to explore with colleagues what variation exists in the lives of humans and dogs in diverse Pacific settings. We challenged ourselves with two big questions. First, how could we make sense of these variants in terms of culture: in humans, in dogs, or somehow in both? Second, could we break free of our native anthropocentrism and begin to get at canine perspectives? We documented culturally diverse human and canine communities. Our efforts to theoretically account for this diversity culminated for me in realizing the concept of human–canine interspecies culture.

Conclusion

Dogs and humans shape one another in diverse ways by expressing cultural information that arises from their interaction according to their own changing capabilities and knowledge. Pacific Oceania's continental landmasses, high islands, and tiny atolls are separated by open seas that, from dogs' earliest arrivals in Australia and New Guinea, have been traversable only by using culture in the form of water-crossing technologies. In the thousands, or hundreds, or tens of years since dogs have found themselves living with and without culturally distinctive humans in numerous societies across the Pacific, dogs have learned and embodied locally and temporally distinctive cultural information even as humans have. Both species, through their mutual engagement, created, developed, enacted, and changed culture in one another, in traditions, and in material and biological effects. This process continues in countless trajectories among canine and human Pacific Islanders and other Earthlings today.

Humans are evidently the senior or dominant partner in interspecies cultural projects with dogs. However, to consider humans sole creators and expressors of these achievements exaggerates human powers and exceptionality. It also underplays the abilities and characteristics of dogs. Individual dogs have lives that include socially engaging with humans, thinking, feeling, learning, and acting in accord with others, though they undoubtedly do all of these in wondrously different ways than that most culturally committed of animals, human beings. Each dog, human, and even fish is an individual with a life story (Balcombe 2010).

The existence and diversity of human–canine interspecies cultures in the Pacific and beyond seems to me amply demonstrated in the chapters of this book: how dogs and humans communicate, their customary activities and expectations, their distinctive contributions to one another's needs, and their antipathies show recurrences and patterns, but are not identical in the cases presented. Both interacting species display social learning and expressions of information once socially learned in their characteristic attitudes, actions, physiques, and territorial ranges. Both interacting species necessarily contribute their parts, according to their different abilities, to remembering, reasoning, enacting, and materializing these cultural representations.

Human societies and their cultures are not cut off from the rest of the living world. They include and mutually engage parasites, prey, pets, predators, beasts of burden, and hunting partners. "Human" societies are not strictly human at all, but include life forms of all taxonomic kingdoms, most obviously our fellow animals with whom we socialize and exchange information. Recognizing and acting on this, as John Knight (2005:1) puts it, involves taking "an interest in animals as subjects rather than objects, in animals as parts of human society rather than just symbols of it, and in human interactions and relationships with animals rather than simply human representations of animals."

Dogs are some of the most obvious nonhuman members of multispecies societies that include humans. Social learning traditions that involve both species have consequences for the acts, bodies, and broader ecosystems in which we are all interdependently enmeshed. We learn from and with one another: "canines and humans collectively construct meaning", Alex Oehler (2018:29) observes, "each actively drawing on the perceptive advantages they recognize in the other's perspective". We are their humans as much as they are our dogs.

References

Altman, Arie, and Alex Mesoudi. 2019. Cultural evolution, gene-culture co-evolution, and cultural niche construction. *Human Ecology* 47: 483–497.

Ameen, Carly, Tatiana R. Feuerborn, Sarah K. Brown, Anna Linderholm, Ardern Hulme-Beaman, Ophélie Lebrasseur, Mikkel-Holger S. Sinding, et al. 2019. Specialized sledge dogs accompanied Inuit dispersal across the North American Arctic. *Proceedings of the Royal Society B* 286(196).

Anderson, Myrdene. 1986. From predator to pet: Social relationships of the Saami reindeer-herding dog. *Central Issues in Anthropology* 6(2): 3–11.

Anderson, Myrdene. 2016. Runne-Beane: Dog herds ethnographer. *Ethnobiology Letters* 72(2): 32–40.

Balcombe, Jonathan. 2010. *Second nature: The inner lives of animals.* New York: Palgrave Macmillan.

Balme, Jane, and Sue O'Connor. 2021. The coming of the dingo. In *The Oxford handbook of the archaeology of Indigenous Australia and New Guinea.* Ian J. McNiven and Bruno David, eds. Pp. 361–380. Oxford: Oxford University Press.

Bamana, Gaby. 2014. Dogs and herders: Mythical kinship, spiritual analogy, and sociality in rural Mongolia. *Sino-Platonic Papers* 245: 1–18.

Beck, Alan M. 2002 [1973]. *The ecology of stray dogs: A study of free-ranging urban animals.* West Lafayette, IN: Purdue University Press.

Bendarik, Robert G. 2008. The domestication of humans. *Anthropologie* 46(1): 1–17.

Boesch, Christophe. 2012. *Wild cultures: A comparison between chimpanzee and human cultures.* Cambridge: Cambridge University Press.

Boyce, James. 2006. Canine revolution: The social and environmental impact of the introduction of the dog to Tasmania. *Environmental History* 11(1): 102–129.

Brumm, Adam. 2021. Dingoes and domestication. *Archaeology in Oceania* 56(1): 17–31.

Burtt, Amanda, and Brandi Bethke, eds. 2020. *Dogs: Archaeology beyond domestication.* Gainesville: University Press of Florida.

Casey, Rachel A., Maria Naj-Oleari, Sarah Campbell, Michael Mendl, and Emily J. Blackwell. 2021. Dogs are more pessimistic if their owners use two or more aversive training methods. *Scientific Reports* 11:19023.

Chao, Sophie. 2022. *In the shadow of the palms: More-than-human becomings in West Papua.* Durham, NC: Duke University Press.

Coppinger, Raymond, and Lorna Coppinger. 2002. *Dogs: A new understanding of canine origin, behavior, and evolution.* Chicago: University of Chicago Press.

Coppinger, Raymond, and Lorna Coppinger. 2016. *What is a dog?* Chicago: The University of Chicago Press.

Coren, Stanley. 2004. *How dogs think: Understanding the canine mind.* New York: Free Press.

Coulter, Kendra. 2018. Challenging subjects: Towards ethnographic analyses of animals. *Journal for the Anthropology of North America* 21(1): 58–71.

Cramb, Justin. 2022. The dogs of Remote Oceania: An archaeological and ethnohistorical view of domestic dog introduction and loss in the South Pacific. *Archaeology in Oceania* 57(1): 28–38.

Cramb, Justin, and Carla S. Hadden. 2020. From land and sea: Overcoming the challenges of directly dating dog remains on Polynesian islands. *Radiocarbon* 62(6): 1–15.

Crawford, Margaret A., John A. Perrone, Clare M. Browne, Catherina L. Chang, Sandra Hopping, and Timothy L. Edwards. 2022. Transitioning from training to testing with scent detection animals: Application to lung cancer detection dogs. *Journal of Veterinary Behavior* 55–56: 23–34.

Cummins, Bryan D. 2013. *Our debt to the dog: How the domestic dog helped shape human societies.* Durham, NC: Carolina Academic Press.

Drenthen, Martin. 2023. Finding common ground with wolves: Interspecies communication in a shared landscape. In *The Wolf.* Pp. 287–294. Martlesham, UK: Boydell & Brewer.

Druzhkova, Anna S., Olaf Thalmann, Vladimir A. Trifonov, Jennifer A. Leonard, Nadezha V. Vorobieva, Nikolai D. Ovodov, Alexander S. Graphodatsky, and Robert K. Wayne. 2013. Ancient DNA analysis affirms the canid from Altai as a primitive dog. *PLoS One* 8(3):e57754.

Dugatkin, Lee Alan, and Lyudmila Trut. 2017. *How to tame a fox (and build a dog): Visionary scientists and a Siberian tale of jump-started evolution.* Chicago: University of Chicago Press.

Dugnoille, Julien. 2019. "I heard a dog cry": More-than-human interrelatedness, ethnicity and zootherapy in South Korean civil society discourse about dog meat consumption. *Ethnography* 20(1): 68–87.

Dwyer, Peter D., and Monica Minnegal. 2021. Relationship between wild-living and village-living dogs in New Guinea. *PNAS* 118(12):e2020432118.

Enquist, Magnus, Stefano Ghirlanda, and Johan Lind. 2023. *The human evolutionary transition: From animal intelligence to culture.* Princeton, NJ: Princeton University Press.

Fitzpatrick, Simon, and Kristin Andrews. 2022. Animal culture and animal welfare. *Philosophy of Science* 89(5): 1104–1113.

Germonpré, Mietje, Martina Galetova, and Mikhail Sablin. 2012 Palaeolithic dog skulls at the Gravettian Predmostí site, the Czech Republic. *Journal of Archaeological Science* 39(1): 184–202.

Germonpré, Mietje, Martina Lázničková-Galetová, Mikail V. Sablin, and Hervé Bocherens. 2023. The advantages of owning a Paleolithic dog. In *Dogs, past and present: An interdisciplinary perspective.* Ivana Fiore and Francesca Lugli, eds. Pp. 63–71. Oxford: Archaeopress.

Golub, Alex. 2005. Shooting Snowy was the toughest job I ever had: The role of dogs in first contact and anthropological theory. Paper presented at "Fashioning Anthropology: Papers in Honor of Gail Kelly" Portland OR, April 15–16. https://www.academia.edu/2748668/Shooting_Snowy_was_the_Toughest_Job_I_Ever_Had_The_Role_of_Dogs_in_First_Contact_and_Anthropolog.

Greig, K., A. Gosling, C. J. Collins, J. Boocock, K. McDonald, D. J. Addison, M. S. Allen, et al. 2018. Complex history of dog (Canis familiaris) origins and translocations in the Pacific revealed by ancient mitogenomes. *Scientific Reports* 8: 9130.

Greig, Karen, James Boocock, Stefan Prost, K. Ann Horsburgh, Chris Jacomb, Richard Walter, and Elizabeth Matisoo-Smith. 2015a. Complete mitochondrial genomes of New Zealand's first dogs. *PLoS One* 10(10):e01328536.

Greig, Karen, Richard Walter, and Elizabeth A. Matisoo-Smith. 2015b. Dogs and people in Southeast Asia and the Pacific. In *The Routledge handbook of bioarchaeology in Southeast Asia and the Pacific Islands.* Marc Oxenham and Hallie Buckley, eds. Pp. 462–482. London: Routledge.

Guillo, Dominique, and Nicolas Claidière. 2020. Do guide dogs have culture? The case of indirect social learning. *Humanities and Social Sciences Communications* 7(31): 1–9.

Hall, Roberta, and Henry S. Sharp, eds. 1978. *Wolf and man: Evolution in parallel.* New York: Academic Press.

Hare, Brian, Michelle Brown, Christina Williamson, and Michael Tomasello. 2002. The domestication of social cognition in dogs. *Science* 298(5598): 1634–1636.

Hare, Brian, and Michael Tomasello. 2005. Human-like social skills in dogs? *Trends in Cognitive Science* 9(9): 439–444.

Hartigan, JohnJr. 2021. Knowing animals: Multispecies ethnography and the scope of anthropology. *American Anthropologist* 123(4): 846–860.

Hiby, E. F., N. J. Rooney, and J. W. S. Bradshaw. 2004. Dog training methods: Their use, effectiveness and interaction with behaviour and welfare. *Animal Welfare* 13(1): 63–69.

Ingold, Tim. 1993. The reindeerman's lasso. In *Technological choices: Transformations in material culture since the Neolithic.* Pierre Lemonnier, ed. Pp. 108–125. London: Routledge.

Janssens, Luc, Liane Giemsch, Ralf Schmitz, Martin Street, and Philippe Crombé Van Dongen. 2018. A new look at an old dog: Bonn-Oberkassel reconsidered. *Journal of Archaeological Science* 92: 126–138.

Jones, Rhys. 1970. Tasmanian Aborigines and dogs. *Mankind* 7(4): 256–271.

Kaminski, Juliane, Juniane Bräuer, Josep Call, and Michael Tomasello. 2009. Domestic dogs are sensitive to a human's perspective. *Behaviour* 146(7): 979–998.

Kirksey, S. Eben, and Stefan Helmreich. 2010. The emergence of multispecies ethnography. *Cultural Anthropology* 25(4): 545–576.

Knight, John. 2005. Introduction. In *Animals in person: Cultural perspectives on human–animal intimacies.* John Knight, ed. Pp. 1–14. Oxford: Berg.

Koungoulos, Loukas, and Melanie Fillios. 2020. Hunting dogs down under? On the Aboriginal use of tame dingoes in dietary game acquisition and its relevance to Australian prehistory. *Journal of Anthropological Archaeology* 58: 101146.

Koungoulos, Loukas George, Jane Balme, and Sue O'Connor. 2023. Dingoes, companions in life and death: The significance of archaeological canid burial practices in Australia. *PLoS One* 18(10):e0286576.

Kulick, Don. 2017. Human–animal communication. *Annual Review of Anthropology* 46: 357–378.

Larson, Gregor, Elinor K. Karlsson, Angela Perri, Matthew T. Webster, Simon Y. W. Ho, Joris Peters, Peter W. Stahl, et al. 2012. Rethinking dog domestication by integrating genetics, archeology, and biogeography. *PNAS* 109(23): 8878–8883.

Lende, Daniel H., and Greg Downey. 2012. The enculturated brain—toward the future. In *The enculturated brain: An introduction to neuroanthropology.* Daniel H. Lende and Greg Downey, eds. Pp. 391–420. Boston: The MIT Press.

Locke, Piers, and Ursula Muenster. 2015. Multispecies ethnography. In *Oxford Bibliographies.* Oxford: Oxford University Press. https://www.oxfordbibliographies.com/view/document/obo-9780199766567/obo-9780199766567-0130.xml.

Lohmann, Roger Ivar. 2010. Introduction: The anthropology of creations. *Anthropological Forum* 20(3): 215–234.

Lohmann, Roger Ivar. 2021. Out-of-body on the happy hunting road: Dialogues between dreaming and culture in Papua New Guinea. In *New directions in the anthropology of dreaming*. Jeannette Mageo and Robin E. Sheriff, eds. Pp. 137–157. London: Routledge.

Lohmann, Roger Ivar. 2022. Human–canine interspecies cultures of Oceania. Paper presented at the Canadian Anthropology Society. Regina, SK, May 11–15.

Losey, Robert J. 2021. Domestication is not an ancient moment of selection for prosociality: Insights from dogs and modern humans. *Journal of Social Archaeology* 22(2): 131–148.

Losey, Robert J., Tatiana Nomokonova, Natalia V. Fedorova, and David G. Anderson. 2020. Domestication as enskilment: Harnessing reindeer in Arctic Siberia. *Journal of Archaeological Method and Theory* 28: 197–231.

Louwaars, Niels P. 2018. Plant breeding and diversity: A troubled relationship? *Euphytica* 214(7): 114.

Lyons, Martyn. 1986. *The totem and the tricolour: A short history of New Caledonia since 1774*. Sydney: New South Wales University Press.

Manne, Tiina, Bruno David, Fiona Petchey, Matthew Leavesley, Georgia Roberts, Katherine Szabó, Chris Urwin, Ian J. McNiven, and Thomas Richards. 2020. How long have dogs been in Melanesia? New evidence from Caution Bay, south coast of Papua New Guinea. *Journal of Archaeological Science: Reports* 30 (2020): 102255.

Marzluff, John M., and Tony Angell. 2005. Cultural Coevolution: How the human bond with crows and ravens extends theory and raises new questions. *Journal of Ecological Anthropology* 9(1): 69–75.

Mech, L. David, and Luc A. A. Janssens. 2021. An assessment of current wolf Canis lupus domestication hypotheses based on wolf ecology and behaviour. *Mammal Review* 52(2): 304–314.

Mikhail, Alan. 2014. *The animal in Ottoman Egypt*. Oxford: Oxford University Press.

Miklósi, Ádám. 2015. *Dog behaviour, evolution, and cognition*. Oxford: Oxford University Press.

Morris, Brian. 1998. *The power of animals: An ethnography*. Oxford: Berg.

Musharbash, Yasmine. 2017. Telling Warlpiri dog stories. *Anthropological Forum* 27(2): 95–113.

Oehler, Alex. 2018. Hunters in their own right: Perspectival sharing in Soiot hunters and their dogs. In *Dogs in the north: Stories of cooperation and co-domestication*. Robert J. Losey, Robert P. Wishart, and Jan Peter Laurens Loovers, eds. Pp. 28–44. London: Routledge.

Oehler, Alex. 2021. Humans and dogs of mountainous inner Asia: Sensory collaboration and personhood. *Human Ecology* 49: 765–778.

Ogden, Laura A., Billy Hall, and Kimiko Tanita. 2013. Animals, plants, people, and things: A review of multispecies ethnography. *Environment and Society: Advances in Research* 4: 5–24.

Orr, Yancey. 2016. Interspecies semiotics and the specter of taboo: The perception and interpretation of dogs and rabies in Bali, Indonesia. *American Anthropologist* 118(1): 67–77.

Parker, Dan, Kylie Soanes, and Stanislav Roudavski. 2022. Interspecies cultures and future design. *Transpositiones* 1(1): 183–236.

Perri, Angela, Tatiana R. Feuerborn, Laurent A. F. Frantz, Gregor Larson, Ripan S. Malhi, David J. Meltzer, and Kelsey E. Witt. 2021. Dog domestication and the dual dispersal of people and dogs into the Americas. *PNAS* 118(6): e2010083118.

Pilley, John W. 2013. Border Collie comprehends sentences containing a prepositional object, verb, and direct object. *Learning and Motivation* 44(4): 229–240.

Qikiqtani Inuit Association. 2014. Qimmiliriniq: Inuit sled dogs in Qikiqtaaluk. In *Qikiqtani truth commission: Thematic reports and special studies 1950–1975*. Toronto: Inhabit Media. https://www.qtcommission.ca/sites/default/files/public/thematic_reports/thematic_reports_english_qimmiliriniq.pdf.

Russell, Edmund. 2018. *Greyhound nation: A coevolutionary history of England, 1200–1900*. Cambridge: Cambridge University Press.

Schleidt, W. M., and M. D. Shalter. 2003. Co-evolution of humans and canids. An alternative view of dog domestication: Homo homini lupus? *Evolution and Cognition* 57(9): 57–72.

Serpell, James. 2016. From paragon to pariah: Cross-cultural perspectives on attitudes to dogs. In *The domestic dog: Its evolution, behavior and interactions with people*. James Serpell, ed. Pp. 300–316. Cambridge: Cambridge University Press.

Shipman, Pat. 2015. *The invaders: How humans and their dogs drove Neanderthals to extinction*. Cambridge, MA: The Belknap Press of Harvard University Press.

Shipman, Pat. 2021. *Our oldest companions: The story of the first dogs*. Cambridge, MA: The Belknap Press of Harvard University Press.

Spottiswoode, Claire N., and Brian M. Wood. 2023. Culturally-determined interspecies communication between humans and honeyguides. *Science* 382(6675): 1155–1158.

Stahl, Peter W. 2016. Old dogs and new tricks: Recent developments in our understanding of the human–dog relationship. *Reviews in Anthropology* 45(1): 51–68.
Stanyukovich, Maria V. 2023. Dog in Philippine life, ritual and creation myths: In a spirit of hunting. In *Dogs, past and present: An interdisciplinary perspective.* Ivana Fiore and Francesca Lugli, eds. Pp. 418–433. Oxford: Archaeopress.
Sueur, Cédric. 2022. Non-human animal cultures, co-cultures and conservation. *Cultural Science* 12(1): 93–102.
Sueur, Cédric, and Michael A. Huffman. 2024. Co-cultures: Exploring interspecies culture among humans and other animals. *Trends in Ecology and Evolution* 39(9): 821–829.
Surbakti, Suriani, Heidi G. Parker, James K. McIntyre, Hendra K. Maury, Kylie M. Cairns, Meagan Selvig, Margaretha Pangau-Adam, et al. 2020. New Guinea highland wild dogs are the original New Guinea singing dogs. *PNAS* 117(39): 24369–24376.
Szabó, Dóra, Daniel S. Mills, Friederike Range, Zsófia Virányi, and Ádám Miklósi. 2017. Is a local sample internationally representative? Reproducibility of four cognitive tests in family dogs across testing sites and breeds. *Animal Cognition* 20: 1019–1033.
Titcomb, Margaret. 1969. *Dog and man in the ancient Pacific with special attention to Hawaii.* Honolulu: Bernice P. Bishop Museum.
Udell, Monique A. R., Nicole R. Dorey, and Clive D. L. Wynne. 2010. What did domestication do to dogs? A new account of dogs' sensitivity to human actions. *Biological Reviews* 85: 327–345.
van der Veen, Marijke. 2014. The materiality of plants: Plant–people entanglements. *World Archaeology* 46(5): 799–812.
van Dooren, Thom, Matthew Chrulew, Myles Oakey, Sam Widin, and Drew Rooke. 2025. Animal cultures at the edge of extinction. *Theory, Culture and Society* 42(2): 3–23.
Wan, Michele, Enikő Kubinyi, Ádám Miklósi, and Frances Champagne. 2009. A cross-cultural comparison of reports by German Shepherd owners in Hungary and the United States of America. *Applied Animal Behaviour Science* 121(3–4): 206–213.
Whiten, Andrew, Christine A. Caldwell, and Alex Mesoudi. 2016. Cultural diffusion in humans and other animals. *Current Opinion in Psychology* 8: 15–21.
Wickler, Stephen. 2004. Modelling colonisation and migration in Micronesia from a zooarchaeological perspective. In *Colonisation, migration, and marginal areas: A zooarchaeological approach.* Keith Dobney, Peter Rowley-Conwy, and Umberto Alarella, eds. Pp. 28–40. Oxford: Oxbow Books.
Williams, Carys Louisa, Silvia Michela Mazzola, Giulio Curone, and Giovanni Quintavalle Pastorino. 2018. What we have lost: Domestic dogs of the ancient South Pacific. *Annual Research and Review in Biology* 25(2): 1–11.
Wynne, Clive D. L. 2019. *Dog is love: Why and how your dog loves you.* Boston: Houghton Mifflin Harcourt.
Zeder, Melinda A. 2012. Pathways to animal domestication. In *Biodiversity in agriculture: Domestication, evolution, and sustainability.* Paul Gepts, Thomas R. Famula, Robert L. Bettinger, Stephen B. Brush, Ardeshir B. Damania, Patrick E. McGuire, and Calvin O. Qualset, eds. Pp. 227–259. Cambridge: Cambridge University Press.

2 What Arapesh Dogs Mean

The Figure of the Talking Dog on New Guinea's Sepik Coast

Lise M. Dobrin

Introduction: What Was (and Wasn't) Said When Piti Caught a Bandicoot

In November 1998, after a year spent conducting linguistic fieldwork in Wautogik Village in the Arapesh "West Coast" region of East Sepik Province, Papua New Guinea (PNG), I took a two-week-long research trip with my husband to another Arapesh dialect area inland along the Sepik Highway. Upon our return to the village, we were delighted when my adoptive mother and brothers informed us that Piti, the dog I had raised from a puppy over the course of the preceding year, had done a lot of bird chasing while we were away. Piti was admired by all who knew him for being an avid hunter, and the reports we received of his bird chasing were consistent with a common village discourse that celebrated his good character. I fed him well, and he was larger and stronger than the other dog from his litter that had grown to adulthood in the village.[1] He followed me everywhere, even to church—and sometimes even hopped up into the bus with me when I was heading to town. He enjoyed the benefits of a flea collar that I had sent to me from the US, so his reddish-brown coat was shiny, with none of the bare patches from scratching that marred the coats of most mangy village dogs.

But over the next few days we began hearing hushed talk from others that while we were away Piti had in fact caught a *mumut* ('bandicoot' in Tok Pisin, Papua New Guinea's creole lingua franca)—a rat-like marsupial that is a staple game of village hunters—and that members of my very own family had cooked and eaten it. I was troubled by the lack of reciprocity that their secrecy entailed. Those to whom I was closest were hiding from me their consumption of meat obtained through my dog's labors, an affront to the intimacy I thought had been achieved over the course of the many warm interactions and intensive exchange relations I had had with my village hosts. Shouldn't the meat from the animal caught by my dog have been shared with me, just as I shared with them the tinned fish and corned beef that I brought back from my trips to town?

DOI: 10.4324/9781003658696-2

And how could they have withheld from me the enjoyment of sharing in the narrative of Piti's exploits? On a previous occasion when Piti caught an animal in the bush and I was present to witness it, I returned home with a detailed report. While Piti and I were out with my maternal cross-cousin Tapinas at her garden he helped her catch a lizard, which she took with my blessing to eat with her family. (I declined my share in it, to my own family's chagrin, but that choice everyone acknowledged to be mine.) When I returned home I dutifully relayed every twist and turn of the story: how the lizard escaped into a shallow stream where Piti lost its scent; how Tapinas managed to recover it from the stream, whack it with her bush knife, and tuck it into her bag; how the lizard roused itself and escaped from the bag; how when she finally killed it and parceled it up she tied the bag extra tightly just in case it should try again to slip out.

Of course, I understood intellectually that in Arapesh country one could never know for sure what transpired unless one saw it for oneself. I also knew why my family would be disinclined to tell me about the bandicoot. Since Piti was mine, those who benefitted from his hunting success were obliged to acknowledge my investment in the dog's growth and strength that enabled him to capture the animal. If I had been with them at the time that obligation would have been acknowledged by all present, I would have declined any share in the bandicoot, and they would have been off the hook. But unless they were prepared to now owe me meat, with the bandicoot long since digested, things were simpler if I just did not know.

Still, my feelings were hurt, so I decided to reveal my cognizance of the situation and express my dismay. Sitting around with my mother and brothers petting the dog one day I lowered my voice and said, *Piti is a very special dog. He knows how to talk. He talks to me at night when we are alone at home. Sometimes he tells me stories. He told me one about a bandicoot.* I knew from their reactions that my indirect message had been received. They gave no indication of being confused or surprised. Instead, they looked around uncomfortably and changed the subject. Piti's hunting exploits in my absence were never overtly mentioned by any of us.

To this day I consider my pointed use of this fantastical story to effectively communicate my displeasure and rebuke my family to be among my most solid accomplishments as a student of Arapesh culture. It is not the judicious use of indirection or speaking an untruth that was significant, for throughout the Pacific those are stock elements in the interactional toolkit of anyone trying to manage their social relations (Myers and Brenneis 1991:14–17). To confront my family members directly would have put us all in a difficult position, implying that I thought they were "greedy" because they had kept the meat for themselves. To make such an accusation would have brought bad feelings out into the open where they could generate conflict, while undermining my implicit claim to being worthy of receiving meat by proving that I was a hopeless social oaf. Furthermore, as elsewhere in PNG, Arapesh villagers are little compelled by the notion that a person's thoughts and intentions should be veridically reflected in their speech.[2] To the extent that I found the departure from truth-telling (on both my family's part and mine) awkward, it was only because I was carrying over my American communicative sensibilities to a different cultural setting where they had little purchase.

What was significant about the story in a specifically Arapesh cultural context was my reference to Piti's ability to speak. Effectively deploying the narrative figure of a talking dog to comment on controversial village events was the ethnographic heart of my interactional achievement. The talking dog was based on a model I had internalized over a year spent looking after Piti and eliciting and transcribing Arapesh villagers' stories, many of which featured dogs that play a special role in relation to humans. In Arapesh narratives, dogs frequently act in the best human-like ways: they intervene constructively at moments of distress, surprise, or potential trouble, helping to stabilize anxious or dangerous situations, and they serve as key conduits in the formation of new social relationships. Because of the common currency of this narrative figure, my suggestion that Piti could speak provoked curiosity, but not confusion. Referencing it softened the impact of my implicit criticism by invoking a knowledge schema that my Arapesh interlocutors and I shared, presupposing rather than undermining (I hoped) the solidarity between us.[3]

I never set out to give ethnographic attention to dogs. But as I worked documenting varieties of the Coastal Arapesh language (see location in Map 2.1) by recording narratives both traditional and emergent, it became evident that many Arapesh stories systematically incorporate dogs into their cast of characters. In some the dogs are incidental, but in others they play a pivotal role in shaping how narrative events unfold. Having been given a puppy to raise and care for in the village as I was transcribing the recordings, I found myself refracting the narratives I was focused on in my linguistic work through my experiences as a participant in the Arapesh dog-human social world.

My aim in this chapter is to introduce the ubiquitous figure of the quasi-human, at times even talking dog, and illustrate its place in Arapesh narrative discourse. The field of human–animal studies is currently less focused on "the question of what dogs and other animals tell humans about ourselves" when they speak than it is on what animals' speech might indicate about the animals themselves (Kulick 2021:8). Nevertheless, anthropomorphic attribution is "an intrinsic human tool for the establishment of connections to people, animals and nature" (Lockwood 2023:105) so can be culturally revealing. In the long tradition of animal studies in anthropology I use Arapesh dog stories critically here as a resource for understanding social relations from a local cultural perspective (see, e.g., Bulmer 1967; Leach 1964; Lévi-Strauss 1964; Mullin 1999; Ohnuki-Tierney 1987; Tambiah 1969; Tuzin 1997). I especially take inspiration from Musharbash (2017), who writes about Warlpiri "dog stories", the stories Warlpiri people

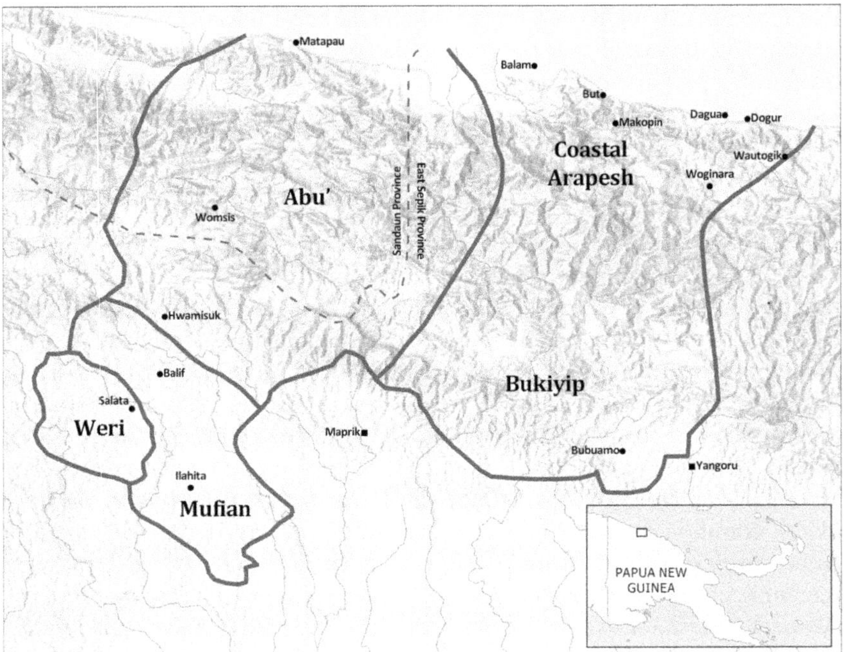

Map 2.1 The Arapesh language area in northern Papua New Guinea. Map by the author.

in Australia tell about dogs, as a way of bringing ethnographic focus to the moral qualities they associate with Aboriginal, as opposed to non-Indigenous *whitefella*, ways of being. Warlpiri talk about dogs suggests to Musharbash a view of being Aboriginal that means living in close proximity with others, serving as non-directive companions and witnesses, and caring for kin.

In a similar way, I interpret the dog as a narrative figure that expresses an Arapesh cultural archetype: a social actor who engages in generous helping. In a helping relation, one exerts agency as one acts in the world, but the benefits of the actions accrue to another. Although helping is a profoundly important Arapesh value, individuals do not always find it easy to subordinate their own goals to those of others. But through what they say and do in Arapesh narratives, dogs show themselves to be good helpers, serving as a model for the kinds of pro-social ways of being that are felt to contribute to a harmonious, fulfilling communal life.

More specifically, in Arapesh narratives dogs often act as 'friends', a culturally particular social relationship which Arapesh value especially because it is elective and because it extends one's lines of connection, influence, and importation across distance and beyond obligation to kin (Fortune 1939; Mead 1938). This concept of friend is the source of the name *Arapesh* that these communities have come to be known by in the ethnographic literature (Fortune 1939:23):

> The Arapesh have no word in their language indicating their entire tribe or their entire country. The word *arapef* means simply friends, and it is their word for their more distant personal connections. This word has been coined in the written form *Arapesh* in order to name their tribe, country, language, and culture.

It is the name by which the people of this area continue to know and speak of themselves today.[4]

Although dogs cannot use human speech, they often communicate using bodily signs and movements that signal a desire to connect. As I will show, this mode of communication aligns dogs with the exciting Arapesh cultural category of newly encountered foreigners who may become "friends", trading partners, and eventually kin, despite not sharing a common language. But although the humans with whom one communicates by non-verbal signs may eventually move into these valued social roles as one learns their languages and they one's own, dogs disappointingly never make the transition. The many Arapesh narratives that dramatize human–dog communication covertly underline a thoroughgoing cultural anticipation that with those who are new, foreign, or other, a fuller, more rewarding sociality may yet come.

The Dog as Ambivalent Alter Ego

In studies that explore the social meaning of dogs in human societies, it is not uncommon to find dogs being treated as intermediate in status between animals and humans. In contrast to other animals, they may be given personal names, referred to with kinship terms, invited into domestic spaces, or buried in marked graves. Dogs' special status as a distinct kind of animal or even "honorary human" (Corbett 2000:58) may be embedded in the structure of the languages people use to speak about them. For example, in Bosavi (Papuan), the noun meaning 'dog' behaves differently from the nouns referring to other kinds of animals; it is its own distinct category at the same level of abstraction (Feld 1998:463). In Orokaiva (Papuan) there is a classifier that applies to nouns referring to all animals other than birds, only the word for 'dog' does not take it (Ira Bashkow, personal communication, January 2023). In the Hare dialect of Slave (Athabaskan), the noun for 'dog' can be pluralized, whereas no other nouns referring to non-humans are marked for plurality (Rice 1989:247). So, dogs may be linguistically constructed in a way that is different from other animals, in the Hare case effectively carrying them across the category boundary that otherwise divides animal from human.

Culturally, the quasi-human status of dogs often coincides with a deep moral ambivalence shaped by culturally specific conceptions of how moral persons should or should not behave. In Steven Feld's (1998) study of Bosavi narrative he says, "Bosavi men routinely blamed their scrawny dogs for their hunting misfortunes. At the same time, men routinely asserted the great prowess of their dogs, and in private they displayed equal amounts of affection or revulsion toward them" (448). Alma Gottlieb (1992:107) describes dogs among Beng people of Côte d'Ivoire as ambivalent alter egos for humans, "reflecting what humans both seek and dread in one another". She analyzes their role in Beng myths as representing "another character who embodies both the hidden hopes and the hidden frustrations of the main character" (100). In shifting between animal and human in myth, Gottlieb argues, the figure of the dog expresses Beng people's positive appreciation for allies, along with their fear of harmful beings like witches who betray their trust. Writing about people living along the American Northwest Coast, Pamela Amoss (1984:293–294) says "the dog served as a ready metaphor in myth and ritual for ambiguous human relationships". Due to their intermediate status, dogs were ideally positioned to perceive and interface with ambiguous feared spirits such as witches, ghosts, and zombie-like "land otters".

The figure of the dog as ambivalent alter ego may be evident not only linguistic-structurally and in symbolic forms like myth and narrative, but in ethnographic descriptions of cultural practices. Olowo Ojoade reports that Zuru, Efik, and Yoruba people of Nigeria appease gods by sacrificing dogs, and not other animals, because they are "the next best victim after man" (Ojoade 1990:221). Savishinsky (1990) describes a Hare community in which humans lived interdependently with a population of sled dogs that outnumbered them four to one. There, dogs in a successful sled team were prized for achieving the human ideal of brotherly solidarity, using their uncanny ability to detect hazards such as fault lines in the ice to protect their owners and helpfully reveal game trails beyond what the human eye could see. But Hare also treated their dogs with impatience and readily disposed of them when they were weak or unwell, a situation in which they hindered swift mobility, thus making them a liability to human well-being.

In the rest of this chapter, I explore ideas about moral persons and their relations with others that emerge from the numerous Coastal Arapesh narratives that prominently feature humans' interaction with dogs.[5] In Arapesh narratives, dogs frequently feature in the role of helpers, acting in ways that Arapesh humans idealize as the height of generous humanity with other humans. But—and here's the rub—dogs rarely live up to that ideal in daily life. So, while as a narrative figure Arapesh dogs are almost entirely admirable in their human-like qualities, a more complete picture shows them to be morally ambivalent in a way that is consistent with the cases discussed above.

The Dog as Helper in Coastal Arapesh Narratives

Nɨmbag^{w} *'Dogs' as an Arapesh Social Role*

In Arapesh country, the behavior of dogs is interpreted through the lens of exchange: like all other social beings, dogs are understood to be both givers and recipients of feeding and care. Including dogs in one's household means committing to feed them alongside the other members of the household. But feeding someone entails an expectation of reciprocity, so a good Arapesh dog should also give something back. What dogs give will ideally be meat, presumably acquired through the assistance they provide to the household in hunting: good dogs are supposed to accompany their owners into the bush where they help them catch animals, such that "the loss of [a] dog might potentially mean the loss of meat" (Feld 1998:464). Hence the archetypal scene of a man out walking in the bush accompanied by his dog (see Figure 2.1). Dogs should allow their human companions to eat any animal they help them catch, i.e., they should "give" the meat to them. This would be construed as an act of reciprocity for the ongoing feeding and care the dogs receive at home.[6] As we saw in the case of Piti and the bandicoot, the core owner–dog exchange pairing can have carryover effects on human–human relations, because when a dog helps someone catch an animal, potential stakeholders include whoever chased or speared the animal, whoever owns the land the animal was caught on, as well as whoever raised the dog such that it had the strength to succeed in hunting.[7] (This will be illustrated again in the story of the founding of Wautogik village that I share below.)

The status of dogs as helpers is made culturally salient in coastal Arapesh villages by the presence of a formal social role known as *nɨmbag*^{w} 'dogs'. As with other social categories, the way this role is structured varies from community to community; in its outlines it is also shared by the

Figure 2.1 Men departing Wautogik village for the bush accompanied by their dogs. Photo by the author.

neighboring Yangoru Boikin (Paul Roscoe, personal communication 2023). In Wautogik village it is built into the clan system. At the top of the hierarchy of clans is the *giɲau*, the leading or "power" clan. This is not a clan name, but a structural position occupied by the "first" clan, *Abahinem*, the clan defined by descent from the eldest son in an idealized ancestral group of brothers (cf. *abahin* 'eldest brother'). It is the responsibility and privilege of the *giɲau* to take the lead in planning village-wide activities such as feasting or warfare—any kind of significant outward-facing political action. Those individuals serving in the *giɲau* role are assisted in their political activities by the *nɨmbag*^{w} 'dogs', or in Tok Pisin *ol dokman*: the "dog" clans. Which of the village's secondary clans comprise the *nɨmbag*^{w} is debated (see Narokobi n.d.; chapter 10), but in their role as *nɨmbag*^{w}, members of those clans are expected to facilitate, support, or, again in Tok Pisin, *harim tok bilong giɲau* 'do as they are directed by the *giɲau*' (a.36).[8] In other words, the *nɨmbag*^{w} do not set the agenda, they follow the *giɲau*'s lead, and in doing so they help them achieve political goals on behalf of the wider community.

In Wautogik, the examples I was offered to help me understand *nɨmbag*^{w} as a category of social structure tended to emphasize a readiness to jump in and fight when called upon (*troimwe hap abus na ol kirap long kaikai* 'if you throw a piece of meat they will jump to go after it and eat it' [eb1.18]). In earlier ethnographic descriptions of Arapesh socioeconomic life based on research in the locality of Alitoa, the description of the dog clans emphasizes their role as contributors in feast-making (Mead 1947:225):

> The man who undertakes to organize any feast … must assure himself of adequate and reliable help from many others. [One way in which he does this is] by establishing formal ties with a group of members of the opposite dual division (ideally), who will formally undertake to help him, and will formally participate in the honors of the feast. The principal of such a feast is called the "trunk" (*bauwanag*), and the men whose help he formally requests and secures are called the "dogs" (*nubag*).

In either interpretation, we find that dogs are serving as a prominent model for a typified assistant, helper, or follower role in the community's dual organization: the efforts of those who occupy this role help their superordinate partners succeed.

What Dogs Do and Say in Arapesh Stories

Traditional Arapesh narratives often make a point of mentioning that key characters are feeding and caring for their dogs. An example of this can be found in a story I was told by John Nasario in the village of

Umanep (formerly an independent locality; today a section of the composite village Woginara 1). The story tells of a man who leaves his home in the Yangoru area and whose exploits lead to the peopling of Walis Island, off the New Guinea north coast in the Bismarck Sea. In an act of anticipatory resentment for his wife's disappointment in him, the protagonist returns with his dogs from hunting in the bush having caught no meat but hands his wife a parceled up "egg", which is in fact his own excrement wrapped in a leaf. She boils a pot of water and begins cooking the "egg". While she is occupied with the cooking, he prepares to leave before she discovers his disgusting deed. He goes inside and packs his bag (Dobrin n.d.: g.80):[9]

> *Əriə nəkita nawic urupat nəhɨri kokwiʃ, fopelaiʃ, nɨmbatiʃ. Nəhɨri nɨmbatiʃ kokwiʃ, ənəniʃ biyeʃ, nɨmbag^{w} ogwogwiʃ biyeʃ. Neraʃ nənaki neʃuruk ʃakɨs, ərə napwe. Napwe nəne wetimok^{w}.*
>
> And so he got up, went into the house, and got four yams. He got four yams, two for himself, and two for his two dogs. He got them and put them in his bag and stayed. He stayed and waited for his wife to finish cooking.

A very similar description of a person's concern for the needs of their dog can be found in the story of the founding of Wautogik village. Meliawi, a man from the Boikin village of Kwangen (just across the language boundary to the east of Wautogik), accidentally wanders into the homestead of an Arapesh stranger, Yəhələgɨr. But instead of killing Meliawi, as might have been expected, Yəhələgɨr kindly offers him some of the wild tree yams that he was cooking. For each wild tree yam Meliawi eats, he tucks one into his basket for what he describes to Yəhələgɨr as his "puppy", who he had left hidden away (along with his son) in a nearby cave. I recorded several versions of this story; the following excerpt is from the telling by Arnold Watiem (Dobrin n.d.:g.11–22):

> *Nagɨrɨk urukwip nakən biep. Əriə ənən nawɔk ənom. Nati nawɔk ənom, ənom nəmubukɨk. Nəmubukɨk arabʉr rabis nəna natik a jɨrɨg, nəbuk ənom copu namɨndək ərə nariken: "Ɲawɔk ənom ɲəbuk ənom—marec?" Əriə naka, "Eikih^{w} nɨmbatih^{w} h^{w}or gani bʉribʉdik^{w}. Ganda bʉribʉdik^{w} Nɨmberəber." Əriə naka, "Iorait. Kare ɲɨrakeci. Ɲɨna ba ɲɨraət nɨmbat. Gi wab. Mucuh. Ba nuhut, nuhut əriə ɲɨna ɲɨminəh." Əriə naka, "Orait." Nəkita, nahɨr urukwip, urukwip biep nasuep, nowar, nowar gani bʉribʉdik^{w}.*
>
> Yəhələgɨr took wild tree yams from the fire and gave Meliawi two. Meliawi ate one and put another away to his side. Yəhələgɨr saw him do this again and said to him, "You eat one and you put one away. Why?" Meliawi said, "My puppy is in a cave over there at Nɨmberəber." Yəhələgɨr said, "That's ok, go get your dog and bring it back.

> Night is falling. You'll sleep with us here. You can go tomorrow during the daytime." Meliawi agreed. He got up, carrying the two yams, and went down to the cave.

From an Arapesh standpoint, social relations are always two-way. If people are caring for their dogs, their dogs should in some way reciprocate. Narrative representations of dog–human exchange conform to this: Dogs sometimes help their human companions when they are in trouble, and their actions keep them from harm. What is often emphasized is dogs' human-like qualities in acting as partners to humans in such ways.

In Dogur village I recorded Christopher Nɨbekop narrating the tale of Saltep, the woman who is credited with introducing yams, a valued food, to the Arapesh coast (Dobrin n.d.: g.36–40). Saltep arrived in Dogur after being rejected everywhere else she went because of her ugly appearance: her body was covered all over with bumps. In the story, an ancestral *masalai* (the Tok Pisin term for a spirit associated with the land; see Slone 2001:xxv-xxvii) by the name of Bɨrɨmin is living at the beach at Dogur with his son and his dog. Bɨrɨmin and his son catch fish in the sea, but the only staple food they had to eat it with was the wood chips left over from canoe-building. When Saltep arrives, they agree to let her stay with them if she promises to help cook for them, since they are always very hungry when they come back from a day out fishing. As they paddle back to shore after her first day as part of their household, they find she has transformed into a beautiful, smooth-skinned young woman. Where did the bumps go? They were the yams, under her skin, which she had now cooked so they could start eating "real food" instead of wood chips. But when the man and his son see the yams they are afraid to eat them:

> *Na hatəkumec ənec nɨmbat pastaim. "Nogut kwəgiagimou ufəh ba ugak." Kwatɨkumom ənombi haka—haka nɨmbat. Hati nɨmbat tafəh, teyarik hənɨp gen. Na hakərip kopaici hatɨkec haka, "Ati wərigɨn aka." Nani niganin hafəh. Hafəh həminek haka, "Oouu, em wərigɨn tru əgɨndə. Owou wawə yoweifi ənefidenef."*
>
> So first they gave some to their dog. "What if she's trying to trick us?" they thought. "What if we eat them and we die?" So, they took out some yams and gave them to the dog. They watched the dog eat them, and it responded by wagging its tail. They said to each other, "I think it's food." The man and his son ate it. They tasted it and said, "Now this is real food. Before the two of us were eating bad stuff."

In other words, the dog not only signals that the food is safe but indicates that it likes it by wagging its tail. Even though Bɨrɨmin's dog cannot speak, it communicates with Bɨrɨmin and his son, giving them the information they need to keep them from harm.[10]

But sometimes dogs do talk, and when doing so will enable them to provide the kind of help their human companions need; they may even transform into people. The most elaborate narrative about this that I recorded was shared with me by Antonia Guaigu, married into Wautogik from the village of Bogumatai on the Arapesh–Boikin border. It was not a traditional narrative, but a report of events that transpired during the Second World War. The war was a crisis that left no patch of the Arapesh world undisturbed. People in the coastal Arapesh villages either fled inland across the Torricelli mountains where they were taken in by distant friends or even by strangers, or they went down to areas of the coast that were under the control of Australian forces. According to Antonia, who was a young girl at the time, at some point during the war the Australians notified the people of Bogumatai that they should evacuate to the coast. But one of Antonia's grandmothers was too old and infirm to make the trip, so she remained in the village alone, with only her dog as a companion. The elderly woman's family set her up with as much food, firewood, and water as they could gather, and they promised to come back for her as soon as the war ended. As time passed, the old woman's supplies began to dwindle. And then something remarkable happened:

> *Ədiə nɨmbat tati wak—owou wɔrigɨn wak, abør, ɲɨgas, catɨh. Wak, okwokit nɨmbat arəgokwit təkitək adia tətanum okokum arəmatokiɲ batowiɲ, yangpelaikʷ arəmatok. Ədiə okwok kokita kwotaumum abo arəmatok morim, ino abo nɨmbat, arəmatok. Əriə kwona kwɔr ɲɨgas, kwɔr əbør, kwona kwɔtep wɔrigɨn ga nɨmbarigəs, em yain–yamo cenekec cakuticɨkɨk cena. Kwonaki ədiə kwani babokekʷ kwapo wəne wɔrigɨn wagɨnəh wapwe. Okwok kwecuh wɔbʉrəb ərɨgəb, okwok kwecuh wɔbʉrəb ərɨgəb, ino abo nɨmbat morim. Okwok arəmatok. Ageɲihʷ hʷapweɲok adiə kwecuh.*
>
> The dog saw that things were running out: no more food, water, or firewood. And so the dog, which was female, turned itself into a young girl and became like a daughter. She was no longer a dog but stood up like a human. She would go break sticks for firewood, fetch water, and go gather food from the gardens that her family had planted before they left. She would come back to the grandmother and the two of them would cook and eat together. One would sleep on one side of the fire, the other would sleep on the opposite side. She was not like a dog, she was a young woman with skin.

When the old woman's family finally returned, they expected to find her dead, but with the dog's help she had survived:

> *Cohʷar ədiə koka, "Ei yapwe." Hanaki hatəgɨrɨri waka, ənəm habo ərit, waka hawi hatɨkok hatəgɨr hokwem ərit əriə kworeh. Kwakəripon "Niganin" kwokana, "Ətɨdə ino nɨmbat. Arəmatokwiɲ batowiɲ*

> *kwənamwei kwapwe. Kotep wɔrigɨn, apig, kwɔr ɲɨgas, abør wacəh wapwe. Okwok copunek ba kuni ei upwe na ərit ətɨdə ədiə pɨsah ohwo bioh—ohwo bioh ubʉh."*
>
> They called out and she answered: "I'm here." The men wanted to put her on a stretcher to carry her back down to the coast, but she began to cry. She said, "My son, this is no dog. This is a girl who stayed with me. She went and found food, cut firewood, and got water. The two of us ate and lived together. If you put me on a stretcher you have to put her on too. The two of us will go down together."

The grandmother explained that when the girl had heard them call out, she turned back into a dog. So, they put the dog on the stretcher alongside her and carried the two of them down to the place where the family had settled.

According to Antonia, this was not as special a situation as it might seem. Before the Christian era when a new order was established that pushed the non-human world to the periphery, dogs were so much like people that they even used to talk. In fact, she said, her parents used to tell her you shouldn't hit a dog, because a dog is really a person, and if you stay on good terms with it you may later benefit from its help. She said:

> *Bipo tru ol tumbuna ol tok noken paitim dok. Dok em man ino save toktok. Paitim em bai yu harim singaut naka "wak wak wa" na em singaut em man ol itok tambu long ol no save paitim dog bipo ol tumbuna bilong mipela ol papa tu. Ol save was, olsem bikpela ol papmama bilong ol, tumbuna bilong ol, ol save tokim olsem dok em iman. Taim yu wanpela istap bai em ken tanim olsem na lukautim yu.*
>
> Long ago our grandparents said you shouldn't hit a dog. A dog is a person, they just don't talk. If you hit a dog you'll hear it cry, "wah, wah, wah," it'll cry like a person. Our grandparents and parents said that too, and their ancestors before them, they said a dog is a person. When you're alone and need help it can change form and take care of you.

That dogs formerly could speak was also emphasized in the story discussed earlier by John Nasario of Umanep about the peopling of Walis Island. Having left his wife, the protagonist Gironua was out in the bush around Yangoru with his dogs when a storm arose. Gironua and the dogs hid in a hollow log, which got washed downriver and ended up on the shore of Walis Island. When they got out on the beach, the dogs tasted the salt water and exclaimed to Gironua how delicious it was. During his telling John interrupted himself to explain to me as an aside that this was because *seiwɔ nubag^{w} gweyagureh boraɲ*: 'it used to be that dogs spoke'.

Listening to Antonia's story about the dog that became a girl led to an extended conversation about the kinds of helpful acts dogs are capable of. Antonia's husband Joe Guiagu explained to me that:

> *Sampela taim ol man igo na iolsem lus long bus o kain olsem. O lus lo bus o yu bagarap o yu go antap long diwai na ipundaun brukim bun o wanem kain—painim accident long bus na nogat man long helpim em long go long ples, ol dok yet iken karim em igo long ples.*
>
> Sometimes people will go into the bush and get lost, or get injured, or fall out of a tree and break a bone, or something like that. They'll have an accident off in the bush where there's no one to help them get home. In that situation dogs can help by carrying the person back to the village.

Struggling to imagine what it might look like for a dog to carry someone, I asked, "You mean like a horse? Like the dog becomes the person's horse?" "Yes," Joe said. "Or like a person. Like people they'll carry you to the village and then turn back into a dog. Dogs do this." In other words, Joe was saying, it's *not* like being carried on the back of a horse. The dog will carry you the way a person would.

The conversation then shifted to a recent situation in which something along these lines had happened. My village mother Scola Sonin and her then twelve-year-old son Elijah joined Joe and Antonia Guaigu in telling me the story. I present their group narrative in a compiled form in English translation below to make the events they related easier to follow:

> It was a brother of Kobu. He was along the road that goes up the mountain to Woginara. He had fallen down and couldn't walk. He had been chasing a pig and it turned around and attacked him. It bit him repeatedly, leaving him for dead. Well, his two dogs saw that he couldn't walk, so they made a stretcher and carried him all the way down to the main coastal road and left him near the village. When the villagers heard Kobu call out they ran to look and there he was. They took him to the hospital and came back and told his wife. So, the dogs looked after him.

While dogs might talk, or at least try to communicate, people don't always listen to them. Antonia spoke about a relative of hers who collapsed and died down on the coast. When the dog saw what had happened, it went to look for the collapsed person's son and jumped up on him crying. But nobody paid attention. If they had paid attention, Antonia says, they would have found him sooner while his body was still good. The dog tried repeatedly to tell them, but the people refused to listen. And so, he remained there for a long time. By the time they discovered him and brought him back his body had begun to decompose.

But when dogs try to communicate, those who listen to them are rewarded. This is illustrated in a *sap*, a traditional tale, that Antonia told me in which a dog features as hero. The action centers around an unnamed village populated only by women. One day, the women went out and cut bamboo.

Each one returned home with a really nice stick of bamboo. One of the women got a crooked bamboo, but she seemed to deserve it, as her body was covered with the snakeskin-like peeling fungal infection *grile*, making her ugly in the other women's eyes. When the women later broke open their bamboo sticks, each one found a man inside and eagerly claimed him for her husband. But when they saw that the man who came out of the crooked bamboo was exceedingly handsome, the other women became jealous. It was unfair that the ugly woman with *grile* skin should get such an attractive man! But she refused to give him up, and he refused to leave her.

Time passed, and the newly formed couples now had children. But the women remained resentful of the *grile* woman's undeserved good fortune in getting and keeping the most handsome man. So, they decided to punish her. They invited her to join them in fishing for river shrimp. While at the river, they tricked her into diving under the overhang at the river's edge where the dirt had eroded away. While she was down under the water, the other women gathered on the land above her and stomped their feet until the dirt broke and fell in upon her, trapping her under the overhang.

They went back to the village, leaving her there to drown. Only her dog stayed behind. The dog began to dig. It dug and dug, crying for its grandmother. (A dog's owner is idiomatically referred to as its "parent" or "grandparent" in Arapesh irrespective of the person's age.) But it was unable to get her body free.

As one woman after another came back to the village, they encountered the *grile* woman's husband along the road, holding their infant son. When he asked them about his wife, one woman after another said she was coming along behind, or that she had just stopped to gather leaves and should be along shortly. But by the middle of the night she still hadn't returned.

And then her dog came home.

Okwokit nɨmbat tənaki, yaken napwe nati nɨmbat əmənəb buuu! Əmənəb babʉh nabɨrək ə nabɨrək əmənəb ətɨb. Tenaki, teyapurək tato, takihi təne gɨgɨremon yaken təbʉsaurikən teyapɨrok tato tereh əriə tatəgɨr. Tənaki teyapɨrok tato təbʉsauriken tereh tatəgɨr. Nati wak ədiə naka, "Ei!" Nəkita nəhɨr nɨganin nasuon, na nagipecə nɨmbat nənak. Nagipwecə nɨmbat nəna nəna nəna nəna nəna nəna catəgɨr ga wɔrɨb, nati nɨmbat tato takih tapwe tokwarec—tarək. Tarək tarək əriə nati-kugu amakeikigəs əyəs. Irohokwikigəs əyas satəgɨr seir. Natik ədiə naka, "Oo, wagudɨr əmənəb becapumok." Ədiə nəbu nɨganin nakɨs nabʉh nasupum nɨmbat narək nabʉro natom, nakih neyetu nagudɨr hɨr-ɨpih^{w} *h*^{w}*abʉh narək a nabʉh natɨk irohokwik*^{w}*, natɨk irohokwik kwatowi nakihi nakurupok.*

And when it came the husband saw that it was covered all over in mud. The dog came and jumped up on the man on its hind legs. It showed that it was glad to see him, scratching, jumping up on him and crying, and then it went outside. The man recognized that the dog

> was trying to tell him something. So, he got up, picked up his son, and carried him out after the dog. The dog led him back to the river. When they got there, the dog began to dig. It dug and dug until the man saw the woman's legs emerge from the ground. And then he understood: "Oh I see, those women smashed in the ground and buried her in there." He put down the child, grabbed a stick, and helped the dog dig. Together they retrieved his wife's body, brought her out, and washed the mud off her.

Distraught, the man had no desire to return to the village. He found a spot in the bush where he could bury his wife. He built himself a shelter and made a home with his son and their dog there.

Some years later, when the son had grown, a coconut palm grew out of the ground right where the wife's head was buried. Eventually the palm carried fruit, and coconuts began falling to the ground. Unsure of what they were, the man took one, broke it open, poured its liquid into a limbum palm sheath container, and gave it to the dog to drink.

> *Nɨmbat tәna tawɔ әbør, nɨmbat tabørәh, teminә әligәbør әtәbør. Na tәne gɨgɨrem yaken takana, "Wa, әbør wɔsik." Natik, әdiә ata nәhɨr nongotep na nawɔr oub. Nawɔrәb, nadɨmәb nәbubuk nati nɨmbat tabʉh tayatәb. Əriә naka, "Oo, wak. Nɨmbat ino tagәk." Cecuh wab, ruahaip, nati nɨmbat ino tagәk. Naka, "Oo. Ating yopuci muguhɔs." Ənәn adai nadi әnәb nabørәh. Na nabørәh nәkә tokwonibør nɨganin nabørәh, omom hawɔr oub habøh, әriә "Wa," nakana. "Wa, yopuci munguhәs ecɨdәk." Adiә, nani nɨganin napwe, әdiә hapwe, әdiә apә nomokәdai әpәkiɲ sas, olsem caka abʉdә arәpeɲig børәg— oub, ohok^{w} әrәpeɲig børәg.*
>
> The dog drank the coconut water and found it tasted sweet. It was happy and said to the father, "Hey, this coconut water is ok." He saw this, broke open another coconut, carved out the meat and gave it to the dog, and the dog ate it all up. He said, "Ok, the dog didn't die." The next morning, he saw that the dog still had not died. He said, "Ok, I think this is good stuff." Then he took some and ate it himself. He gave some to his son and they both ate it. The man said, "No, this really is good stuff." So, the man and his son stayed living there, and we now have this story. They say that coconuts are a person's head.

In all these stories told to me by Arapesh people, dogs are presented as reliable, intimate friends. In the narrative moments that feature them dogs act in supportive ways, building a stretcher and carrying an injured person to safety, chopping firewood and carrying water for an old woman left alone, trying a new unknown food so people can discover that it is safe for them to eat, notifying someone that things are amiss with a kinsman. In their role as helpers, dogs communicate effectively with their human partners, whether they use speech, 'cry' (Arapesh *-reh*), or "say things" non-verbally. (In Antonia's last story explaining the origin of coconut palms, the dog's approval of the coconut water is formulated as a quote, i.e., as literally spoken.) Although it is less apparent in translation than it is in the Arapesh source material, as dogs "stand up" in the sense of being reliable friends, they also stand up physically, becoming more vertical in comportment, lifting their front legs off the ground as suggested by the choice of verbs (*-to* 'go up', *-yapɨrok* 'jump up') and imagery (they carry like a person, not like a horse). The Arapesh narrative figure of the dog is quasi-human in its role as helper.

Arapesh Dogs as Vectors for New Human Relationships

In Arapesh narratives, dogs' participation in the social order goes beyond providing support to the humans they already have relationships with. They also help humans spearhead new ones. In Arapesh, relationships across social distance are culturally constructed as *yegwih* 'roads', pathways that connect people beyond localities (Dobrin and Bashkow 2006; Fortune 1939; Mead 1935, 1938, 1947, 1961[1937]). Roads are concrete physical paths that link places—and the people who live in them—to one another, but they also refer metaphorically to lines of interpersonal connection. "Road friendships" are treasured in Arapesh country, and their origin stories are recounted with pride. The great accomplishment profiled in these stories is not just the building of new friendships, but the building of new friendships with previously unknown people who speak another language (Dobrin 2021). In Arapesh narratives, when newly introduced interlocutors do not (yet) share a common language, their initial communication can only take place nonverbally through use of facial expressions and gestures. Dramatizing that speaking was impossible because there is no shared language is a sign that significant social distance has been traversed. Having had to communicate nonverbally contributes to the value of the friendship that results, because it indicates that the obstacles to achieving it were high.

The building of the *yaimbon hula* or 'wallaby friendship' road is one such episode in the story of the founding of Wautogik village. The road connects Wautogik villagers from the Abahinem clan, who descend from the village ur-ancestor Yəhələgɨr, with Boikin speakers from the neighboring community of Walanduom (Dobrin n.d.: g.47–56, Narokobi n.d.; chapter 5). The story begins with Yəhələgɨr sitting along the river at the homestead he had recently established with his wife, carving a wooden plate. The scene is suddenly disrupted by a dog chasing a wallaby. Yəhələgɨr gets up and kills the wallaby. Here is how the Arapesh intellectual Bernard Narokobi narrates the ensuing events (in English) in his written *History of Wautogig Village* (Narokobi n.d.; chapter 5):

> [The dog] raced through the river followed closely by two hunters, Heimeri and Kapowia. When they saw Yəhələgɨr and his wife, they stopped. By the ancient laws of hunting, Yəhələgɨr knew the wallaby belonged to the hunters and their dog. So Yəhələgɨr stood up, picked up the dead wallaby, and offered it to the hunters. It was their property, even now that it had been captured, as they and their dog had started the chase and it had come from their land. Heimeri and Kapowia made sign language to Yəhələgɨr and his wife that they could keep the prey. They had killed it. They could keep it.
>
> The two hunters were from Walanduom nation, where the Sausa or Boikin language was spoken. Their sphere of influence extended over parts of the northern banks of Urieb river. They were complete strangers to Yəhələgɨr who was a Buki or Arapesh speaker with origins from Kumunim to the west.
>
> Using sign language, the two hunters advised Yəhələgɨr to keep the wallaby since he had killed it. Again, by sign language, they agreed they would have another meeting at the same place. To aid their memories, they cut midribs of branches of a wild palm tree and indicated that for every day that passed, they would each remove a leaf until they came to the last leaf. When the last leaf was reached, they would reassemble at the beach at the riverside and exchange gifts. They parted ways. On the appointed day they met again and exchanged gifts. Yəhələgɨr gave Heimeri and Kapowia some wooden bowls. Heimeri and Kapowia gave coconut and yam seed to Yəhələgɨr to plant.
>
> In this way, the two nations and two peoples established diplomatic links with each other. Though they spoke different languages and belonged to different communities, the two became allies and close friends. Even to this day, the Walanduoms and the Wautogigem have a close friendship.

A peaceful encounter like this one between people from distant social groups is highly reportable because it cannot be taken for granted that the outcome of such a meeting will be friendship rather than one or the other party's violent demise. The formulaic question regularly asked at such a moment gives some indication of the stakes: *Ne ərəpeɲ aka segəben?* 'Are you a person/friend or a ghost?' Indeed, in all three tellings of the *yaimbon hula* story that I recorded, when Yəhələgɨr and the hunters first meet, a variant of this phrase occurs. In one telling, the narrator uses the term *hərin* 'enemy' in place of *segəben* 'ghost' before correcting himself. Communication in these encounters is described as proceeding in a very specific way: the participants *caku yeməg wis etiɲ* 'they use only facial expressions and hand gestures', emphasizing that the encounter crosses a language boundary.

Clemen Hayin's telling of the encounter between Yəhələgɨr and Meliawi, a different version of which was discussed earlier, likewise illustrates these patterns (Dobrin n.d.: g.47–56). Before approaching Yəhələgɨr, Meliawi says to his son and his dog (who were hidden away in a cave), *Ipə pɨpwe əgɨndək, na ei ito iti gandə arəpeɲ aka səgəben.* 'You guys stay here, and I will go up there and see whether that is a person or a ghost.' Meliawi does this fearing that he could be killed: *nongut nei o* 'he might kill me'. When Yəhələgɨr detects someone coming from the sounds of Meliawi's footsteps breaking twigs on the forest floor, he calls out to the stranger:

> *Naku yeməg etiɲ, narikən, naka, naku wis nakəripon naka, "Yowi." Na natəgɨrumon hapwe.*
>
> [Yəhələgɨr] just used facial expressions to ask him, he used his hands and said, "Come here." And [Meliawi] emerged [from the forest] and they sat down together.

Note that the verbs "ask" and "say" in the gloss above faithfully translate the Arapesh; the encounter is presented as one in which the people who meet are *speaking* when they use their face and hands.[11]

Meliawi's commitment to his son and his dog indirectly contribute to the development of the road connecting Wautogik and Kwangen; recall that Meliawi visibly diverts some of the wild tree yams he was given by Yəhələgɨr so he could feed them, and this action reveals his companions' presence. The dog's help in hunting also leads to the encounter between Yəhələgɨr and the two men from Walanduom, Heimeri and Kapowia, ultimately resulting in the *yaimbon hula* road. In both these stories, the dog is a character on the scene, and its actions have some bearing on what happens, though it is not a direct actor in establishing the road.

However, in a story told to me by Wilson Marai of Balam village (Dobrin n.d.: r2.4–11), the dog is the road's immediate cause. In Arapesh terms, we would say that the dog is the agent who "opens" the road. In this most spectacular instance of the road-building narrative genre, the dog is a named hero, Abɨləminə. Abɨləminə hails from Ɲamikum, a village far inland near Maprik where the Abelam language is spoken. Fed up with the uncaring treatment it gets from its human "mother" who sometimes beats it and only erratically feeds it, Abɨləminə runs away. The dog treks up and down the Prince Alexander mountains via a long series of named locations, ending up at Balam village on the coast. There the dog finds that the people are friendly and the sea is a source of delicious salt. When Abɨləminə signals to its Balam hosts that it is ready to leave, they load the dog up with gifts to carry home. Sometime later Abɨləminə returns, accompanied by an entourage from Ɲamikum who bring with them gifts in return. This two-way exchange of gifts is what formally establishes the road.

Abɨləminə's actions are guided only by its own will. The dog's agency is highlighted in the description of what the dog does when it first sets out: *tarɨtowi bəlugis, ino tɨne saveum yah, tenaki long atətəneɲ laik* 'it goes up into the bush not following a road; it goes wherever it likes'.[12] Abɨləminə's successful nonverbal communication with the Balam villagers about where it comes from and what it wants parallels the successful use of gestures in the previous road-building story:

> *Apəkənəm babokenihem, pepehem, hatɨrɨt na haka, "Atɨdə nubat tənaki agɨnɨm? Atə nubat əmiənət?" Orait eʃeʃ ʃadikət namɨndə na atətənəp hadup pogalihəg. Hadup atətənop takɨripeʃ. Orait ʃalikəgət ʃaka, "Ɲə ɲənaki agɨnɨm?" Na taka, "Ei yenaki ganik."*
>
> Our grandfathers and great-grandfathers saw the dog and said, "Where did this dog come from? Whose dog is it?" They asked it and it wagged its tail. It spoke to them with its tail. They asked, "Where did you come from?" And the dog said, "I came from over there."

After staying in Balam for a week and a day, the dog began to make it known to the villagers that it was ready to leave by "telling" them so.

> *Orait bihain nau taleh. Taleh takɨripeʃ. Takɨripeʃ namɨdə atət taka ka tɨtanum Maprik. Tɨtanum gani Ɲamikum. Orait taleh takɨripeʃ, tena taleh, tagurukweʃ, tawiʃ aiyagəs, tawiʃ alegeguhʷ, təna tatəgɨr gani tokwɔn, tətanum təna tatəgɨr gani yah. Taka tɨtanum dəhɨdə yah nɨmbokɨhi tene tenamərim. Orait tətanum tatem təna. Tətanomi takɨripeʃ ʃaka, "Oo." Debeiʃi ʃene save ting—urukis bɨrarəs nau, "O, atət nubat taka ka tɨtanəm."*
>
> The dog began to cry. It cried and told them that it wanted to go back to Maprik, to Ɲamikum. It cried and told them. It nipped at them, went under and around their legs, went a little way out on the road and then came back. The dog came back and told them, and they said, "Oh, ok." The big men understood. They thought, "Ok, this dog wants to go home."

When Abɨləminə arrived back in Ɲamikum, the gifts it bore made quite an impression. The villagers loved the salt water, but they had no idea where it had come from, because they had had no previous contact with the coast. A month later, Abɨləminə began signaling that it wanted to go back to Balam.

> *Ənəh numinəh nau wabətig, tənak na talih. Talih na takɨripeʃ. Talih, təna tagurukeʃ taka tətanəm tagiməri yah. Wota ʃigikuməti uwe, wota tətanum. Təna tagɨrukʷeʃ, telaweʃi gen. Tənaki, aligaligah, ʃenaki ʃene save, "Ah! nubat taka tɨtanum!" Orait eʃeʃ nau ganik Ɲamigum, wolaiʃi dəbei-dəbeiʃi ʃəkitak, ʃabreri uʃalɨh ʃaru kakiʃ, ʃaruk saukʷeɲ, ʃaruk nɨgalɨh, əneʃ dɨdakeʃ ʃaʃeh, ʃaʃeruki wantaim. Orait. ʃagikuməri nubat. Nubat tariki. Na eʃeʃ dəbeini wagun ʃagikuməti.*
>
> One day as evening fell, the dog went around telling people. It began going around trying to get them to follow. It went around nipping at them. It went out onto the road. When the people didn't

> follow, it came back. It went and nipped at them trying to get them to come along. After a while the big [senior, important] men got it. "Oh I see, the dog wants to go back!" And so, the big men of Ɲamikum got up and started filling up net bags. Many men filled up net bags with yams, tobacco, taro, and other foods. And then they followed the dog. The dog led, and a long line formed behind it.

As they retraced the dog's original path through the unfamiliar bush, some of the men began to lose heart. They feared they were being led to their death. Still, they stayed with the dog until they arrived in Balam.

> *Orait apəkənəʃ dəbei-dəbeiʃi, bobokenihem na amahem, ʃati dətɨdə nubat: "Ei! dətɨdə nubat tətanəmi gen!" ʃənaki ʃati ol deʃɨdə dəbeini lain, ʃagikəməri nubatəri ʃənaki ʃape. Na eʃeʃ ʃapunek apəkənəʃ ʃapunek ino ʃene saveum eʃeʃineɲ tok ples. Eʃeʃineɲ bəlaɲ, na aʃeʃ ʃapunek ino ʃɨne saveum apəkiniɲ bəlaɲ. ʃape ʃene loguh^{w} meioh. ʃape namɨdə ʃagɨgakum. Orait nau nubat gen təlaueʃ ʃəna ʃarei yous, na ʃarei yous, ʃawɔr anəg bɨrabig. Na ʃarei yous, na ʃawɔr anəg bɨrabig na ʃəlaueʃ naka ʃətanəm. Nubat təlaueʃ ka ʃɨtanəm. ʃena ganik Ɲamikum. Nubat tapweum. Na dou walbim, apəkənəm, hagikəməri dəhɨdə yah, hagikəməri nubat atɨt. Nubat seiwɔk tagwehahəri yah.*
>
> And then our grandfathers and great-grandfathers saw the dog. They said, "Hey! That dog has come back!" And they saw this great line of people who had come following the dog. But they didn't know one another's language. We didn't know those people's language, and they didn't know ours. They spoke to each other using only their hands. They couldn't understand. The dog brought them down to the sea and they filled up salt water and gathered salt-laden chunks of driftwood. The dog brought them and then they all went back to Ɲamikum, the dog with them. And now, those inland people and us, we follow this road that the dog made back and forth. This road that was made by the dog.

Thanks to the actions of Abɨləminə, the story ends with geographically distant people being connected despite previously not knowing one another's languages. This is the valued state that obtains when a new road is opened—a new friendship is established—across a major social divide.

Conclusion

An Arapesh Take on "Transspecies Pidgin"

In Arapesh narratives, the expressive behavior of dogs is systematically construed as "talk". But the efficacy of dogs' talk is limited by a mutual intelligibility boundary, such that both humans and dogs must expend special effort to send and interpret messages across it. The "talk" of dogs exhibits parallels to the nonverbal communication that takes place between humans who share no common language. Dogs walk under and around their human interlocutor's legs; they scratch, jump up, cry, and try to pull their human companions in the direction they want them to go. Analogously, humans must use facial expressions and gestures when they encounter people who speak a different language: *caku yeməg wis etiŋ* 'they use only facial expressions and hand gestures'. Linguistic anthropologist Mara Green (2014) refers to such willingness—indeed eagerness—to communicate across a language boundary "linguistic commensuration". It pulls those who otherwise exist on different sides of a linguistic boundary toward equality, reframing them as social beings of a common kind.

Arapesh are not alone in making the implicit assumption that communication with beings of another kind calls for use of non-symbolic expressive means (Kohn 2007; Nadasdy 2007; Smith 2012). In analyzing the semiotic qualities of speech directed at dogs among the Runa of Ávila, Ecuador, in Upper Amazonia, Eduardo Kohn builds on the model of the pidgin languages that emerge when groups of people with no shared language come together for work or trade. Kohn analyzes Runa speech with dogs as a kind of "transspecies pidgin", a reduced form of language that "incorporates elements of communicative modalities from both human and canine realms" (2007:14). As Kohn explains:

> [Transspecies pidgin] is oriented toward beings whose linguistic capabilities are in question. Although people in Ávila go through great efforts to make their dogs understand human speech, how they communicate with their dogs must also conform to the exigencies of those species that cannot normally understand human speech.

Kohn's main concern is with what Runa have to say to their dogs, whereas Arapesh narratives focus more on what dogs have to say to their human companions. Nevertheless, treating speech across species in the same way as speech with socially distant others—as Arapesh do given the common way these scenarios are treated in their narratives—is enlightening in a different way from what Kohn intends, as it leads us to appreciate not just how human sense-making must reckon with a more-than-human world, but how radically "other" socially distant humans can be. For Arapesh, establishing relations with people who speak a foreign language is not so very different from succeeding in communication with a being of another species. Where there is no shared language, speaking must take place through signs and gestures.[13] Indeed, it seems that it is the dog's radically different communicative nature that makes it such a good ambassador to socially distant humans.

The Real vs. Ideal Arapesh Friend

The dogs that feature in Arapesh stories express an Arapesh archetype: a friend who orients toward others and acts in their interest. In the ideal scenario, an Arapesh dog will be cared for generously by its human companions, and it will reciprocate this care by hunting and providing its human companions with meat. It will "stand up" for them in their moments of uncertainty or need, "talking" in its way, and at times taking human form in order to help them. A dog can even help bring about friendships with distant others, one of the most valued possessions Arapesh people can have. Given the frequent appearance of "talking" dogs in Arapesh narratives, it should be clear now why it wasn't entirely out the question that Piti could have spoken to me about a bandicoot, and why my suggestion to my family that he had done so was not laughed at or rejected.

Perhaps unsurprisingly, the moral positioning of dogs in Arapesh people's real lives is a lot messier than it is in their narratives. Given the chance, most dogs will greedily gobble up any meat they can catch.[14] This can lead to disappointments, like the one I experienced on a visit to Wautogik in 2017 when a neighbor's dog pushed open the door to the house I was staying in and stole the package of precious groceries (including fresh chicken and cheese) that I had left on the table following a hard day's shopping trip to town. Dogs' hunting exploits can lead to social conflicts, such as the muted one I had with my adopted family when Piti caught a bandicoot, or the violent tug-of-war that threatened to ensue when Heimeri and Kapowia's dog chased a wallaby from their lands into Yəhələgɨr's homestead but Yəhələgɨr was the one who killed it. Arapesh villagers are frequently disappointed by their dogs, as they are not nearly the helpful friends people imagine they could be.

I learned this both from Arapesh dog stories and from caring for my own dog, Piti, in Wautogik. Piti was admired for being strong, good at hunting, and unfailingly loyal, qualities that drew frequent comment. Piti was exceptional in seeming to live up to the archetypal ideal. That ideal was also realized in one of the villagers' own dogs, Sembo, whose remarkable qualities drew special recognition. Sembo was introduced to me in a story told in English by Arapesh anthropologist Andrew Moutu in April 2023.[15] Sembo had belonged to Andrew's father. He was a formidable hunter, renowned for having caught many bandicoots, at least three wild pigs, even a cassowary. "It fed the village, that dog". Sembo acted in human-like ways, for example "crying" just as a person would upon catching sight of Andrew when he returned to the village to mourn his father's death. When Sembo died he was given a proper burial. Rather than burying the dog off in the village periphery, as Andrew's family would ordinarily have done, they assigned Sembo a burial site in front of their house. The villagers also refrained from going to their gardens for a day after the dog's death, a precaution against spiritual harm that is normally taken when people die. In doing these things, Andrew explained, the villagers "showed that dog the respect that is accorded only to humans".

Sembo was a helpful companion and a generous exchange partner, highly valued qualities that are too often wanting in relationships between humans. In short, Sembo realized the ideal of a good Arapesh friend. What that means is something we learn from Arapesh narratives that put the model for it on full display in the ubiquitous figure of the quasi-human, at times even talking, dog.

Acknowledgements

Most of the stories excerpted here were collected during my dissertation fieldwork (1997–1999). The research was supported by an NSF Doctoral Dissertation Improvement Grant, a Wenner-Gren Foundation Predoctoral Grant, and a Fulbright-Hays Training Grant for Doctoral Dissertation Research Abroad. Development of the Arapesh Grammar and Digital Language Archive was supported by an NSF/NEH Documenting Endangered Languages grant and an NEH Documenting Endangered Languages fellowship. I am deeply indebted to each one of the Arapesh storytellers whose words I recorded and bring together in this chapter. I am also indebted to many other Arapesh people who have helped me understand their meanings. Among these, I owe a special debt to James Moutu, Giamur Mujuwe, Tony Nindim, Matthew Rahiria, Jacob Sonin, Elijah Sonin, Scola Sonin, Patrick Sunan, and Julius Yehaipim. Thanks to Greg Bablis, Ira Bashkow, Larry Carucci, Eve Danziger, Joseph Gabut, Don Kulick, Roger Lohmann, Emmanuel Narokobi, Vergil Narokobi, Paul "Jim" Roscoe, and Colleen Winkelman for thinking along with me while I was writing this chapter. My work on the topic of Arapesh dogs has benefitted from the feedback of audiences at a University of Virginia Linguistic Anthropology seminar in March 2023, a session on "Language at the Boundary of the Human" held at the annual meeting of the American Anthropological Association in Toronto, CA, in November 2023, and three iterations (in 2021, 2022, and 2023) of the Association for Social Anthropology of Oceania session on "Dogs in the Pacific" ably organized by Roger Lohmann. I received invaluable comments from two anonymous reviewers. Courtney Cox and Drew MacQueen worked with me to create the map. Publishing this chapter provides the occasion to recall with love my canine companions past and present: Cokie, Buffy, Piti, Tati, and Ino, all of them dear friends.

Notes

1 Lohmann (this volume) presents another example of puppies from the same litter growing up differently because they were raised in different ways. This, he argues, is evidence that humans and dogs share a common "interspecies

culture". In Lohmann's example the human contributions to the interspecies culture are both Melanesian, whereas the example here highlights a contrast between the Arapesh village style of dog-rearing and the author's American style, in which dogs are treated like privileged (American) children: fed high quality food, bathed, raised inside a shared living space with humans, provided with preventive medicine, etc.

2 Arapesh village children learn from a very young age that what other people say may or may not be true; for example, in order to distract them from crying or whining for a desired object, children's mothers often say untrue things to them things like, *Stop crying or Auntie Lise will give you an injection!* or *Shh! Do you hear the helicopter? A helicopter is coming and it's going to land right here in front of the house!* Such captivating scenarios, which of course never materialize, effectively quiet children while simultaneously training them that what people say to them cannot be taken at face value (see also Kulick 1992; Slotta 2023).

3 Tannen (2004:401) discusses how in middle class American families, too, dogs and other household pets are strategically used as "resources by which speakers buffer criticism, effect frame shifts, deliver praise, teach values, mediate or avoid conflict, and both reflect and constitute the participants' family identities", although of course Arapesh and American messages and contexts differ.

4 For analysis of the name *Arapesh* as a modern identity category see Bashkow (2020:191–198).

5 There are strong resemblances between what dogs do in Arapesh narratives and the kinds of "affiliative practices" documented in Papua New Guinea Highlanders' folk narratives, where dogs contribute to the discovery of new settlements and use human language to communicate (Stewart Strathern and Strathern 2023 ms.).

6 The "inclusion of dogs into exchange activity" with local people is a thread running through accounts of exploratory patrols in multiple areas of early 20th-century New Guinea (Golub 2005:11).

7 The Arapesh language has an elaborate noun classification system that categorizes nouns based on their final consonants (Dobrin 2012). Only nouns referring to humans are categorized according to their semantics. So, in Arapesh, the noun *nɨmbat* (SG) ~ *nɨmbag^w* (PL) 'dog' unambiguously references a nonhuman referent because it ends in the phonemes /t/ (SG) ~ /g^w/ (PL). My use of the English pronoun "it" to refer to a dog in the singular reflects this Arapesh categorization, as most of the examples from the narratives were originally provided in the Arapesh language. Tok Pisin sources present a different challenge for translation: Tok Pisin does not distinguish between the English personal pronouns "it", "him", and "her", collapsing both humanness and gender into a single singular pronoun *em*. Since I am trying to capture the local cultural perspective expressed in the examples, I do my best to follow the original language in each instance, using "it" where the pronominal marker is clearly referencing the noun "dog". Only where the pronominal reference is overtly marked by the speaker as gendered do I use the English forms "she" or "he". I do this in my opening story about Piti because it is told from my own cultural point of view as an American.

8 References that take the form (letter.number) index materials included in the *Arapesh grammar and digital language archive* (Dobrin n.d.). References follow the format described at http://www.arapesh.org/field_notebooks.php.

9 After consultation with Arapesh interlocutors I have chosen to provide vernacular (or Tok Pisin) excerpts of most narratives as well as their English translations. They are lightly edited for clarity. The Arapesh text is roughly phonemic. The only departures from the standard interpretation of IPA

symbols are *c* and *j*, which I use to represent voiceless and voiced alveopalatal affricates, respectively.

10 The protagonists' willingness to subject their dog to poisoning from the new food suggests that humans' care for their dog may have its limits. But the threat of this happening is more apparent than real, as it is not unusual for a historical narrative told in a Papua New Guinean context to dramatize a panoptic "divergence between the characters' perspectives and that of the story's audience", who well know the yams are not poisonous as they form a regular part of their own diet (Slotta 2023:60).

11 In calling for linguists to broaden the scope of their research to include the many diverse registers and lects used across Amazonia, Epps (2021:15) observes that specially marked "speech forms distinct to the interaction at hand rather than independently associated with either participant" can "create a sort of convivial middle ground, particularly where there is a risk of conflict". The gestural register used in (narrated) encounters like those between Yəhələgɨr and Meliawi illustrates something similar.

12 This is reminiscent of the self-directed "going around" that Katharina Schneider uses as diagnostic of the agency of ocean fish that is mimicked by the movements of Pororan Island fishermen in Bougainville (2013:31).

13 Robbins Burling (1999:334) interprets this kind of non-symbolic "groping" as "a stage of communication that amounts to a sort of pre-pidgin".

14 Greediness is also a theme in Feld's (1998) symbolic analysis of Bosavi dogs.

15 The recording of this narrative is held in the author's active research files. It has yet to be transcribed and archived.

References

Amoss, Pamela. 1984. A little more than kin, and less than kind: The ambiguous Northwest Coast dog. In *The Tsimshian and their neighbors of the North Pacific Coast*. Jay Miller and Carol M. Eastman, eds. Pp. 292–305. Seattle: University of Washington Press.

Bashkow, Ira. 2020. The cultural and historical openness of Bernard Narokobi's 'Melanesian Way'. *The Journal of Pacific History* 55(2): 187–219.

Bulmer, Ralph. 1967. Why is the cassowary not a bird? A problem of zoological taxonomy among the Karam of the New Guinea Highlands. *Man* 2(1): 5–25.

Burling, Robbins. 1999. Motivation, conventionalization, and arbitrariness in the origin of language. In *The origins of language: What non-human primates can tell us*. Barbara J. King, ed. Pp. 307–350. Santa Fe: School of American Research Press.

Corbett, Greville G. 2000. *Number*. Cambridge: Cambridge University Press.

Dobrin, Lise M. 2012. *Concreteness in grammar: The noun class systems of the Arapesh languages*. Stanford Studies in Morphology and the Lexicon. Stanford: CSLI Publications.

Dobrin, Lise M. 2021. *The acquisition of other people's languages: A view from Arapesh country*. Unpublished manuscript, University of Virginia.

Dobrin, Lise M. collector and editor. n.d. *Arapesh grammar and digital language archive*. Institute for Advanced Technology in the Humanities, University of Virginia. http://www.arapesh.org/.

Dobrin, Lise M., and Ira Bashkow. 2006. "Pigs for dance songs": Reo Fortune's empathetic ethnography of the Arapesh roads. In *Histories of Anthropology*

Annual vol. 2. Regna Darnell and Frederic Gleach, eds. Pp. 123–154. Lincoln, NE: University of Nebraska Press.

Epps, Patience. 2021. Diversifying multilingualism: Languages and lects in Amazonia. *International Journal of Bilingualism* 25(4): 1–20.

Feld, Steven. 1998. They repeatedly lick their own things. *Critical Inquiry* 24(2): 445–472.

Fortune, Reo. 1939. Arapesh warfare. *American Anthropologist* 41(1): 22–41.

Golub, Alex. 2005. "Shooting Snowy was the toughest job I ever had": The role of dogs in first contact and anthropological theory. Portland, OR: Fashioning Anthropology: Papers in Honor of Gail Kelly. http://hdl.handle.net/10524/1553.

Gottlieb, Alma. 1992. *Under the kapok tree: Identity and difference in Beng thought*. Bloomington: Indiana University Press.

Green, E. Mara. 2014. Building the tower of Babel: International Sign, linguistic commensuration, and moral orientation. *Language in Society* 43(4): 445–465.

Kohn, Eduardo. 2007. How dogs dream: Amazonian natures and the politics of transspecies engagement. *American Ethnologist* 34(1): 3–24.

Kulick, Don. 1992. *Language shift and cultural reproduction: Socialization, self, and syncretism in a Papua New Guinean village*. Cambridge: Cambridge University Press.

Kulick, Don. 2021. When animals talk back. *Anthropology Now* 13(2): 1–15.

Leach, Edmund. 1964. Anthropological aspects of language: Animal categories and verbal abuse. In *New directions in the study of language*. Eric H. Lenneberg, ed. Pp. 23–63. Cambridge, MA: MIT Press.

Lévi-Strauss, Claude. 1964. *Totemism*. Translated by Rodney Needham. London: Merlin Press.

Lockwood, Randall. 2023. Anthropomorphism and anthropocentrism in the Anthropocene. In *The Routledge international handbook of human-animal interactions and anthrozoology*. Aubrey H. Fine, Megan K. Mueller, Zenithson Y. Ng, Alan M. Beck, and Jose M. Peralta, eds. Pp. 93–108. London: Routledge.

Mead, Margaret. 1935. *Sex and temperament in three primitive societies*. New York: William Morrow.

Mead, Margaret. 1961[1937]. The Arapesh of New Guinea. In *Cooperation and competition among primitive peoples*. Margaret Mead, ed. Pp. 20–50. Boston: Beacon Press.

Mead, Margaret. 1938. *The Mountain Arapesh: I. An importing culture*. Volume 36, Part 3. New York: Anthropological Papers of the American Museum of Natural History.

Mead, Margaret. 1947. *The Mountain Arapesh. III. Socio-economic life*. Volume 40, Part 3. New York: Anthropological Papers of the American Museum of Natural History.

Mullin, Molly H. 1999. Mirrors and windows: Sociocultural studies of human–animal relationships. *Annual Review of Anthropology* 28: 201–224.

Musharbash, Yasmine. 2017. Telling Warlpiri dog stories. *Anthropological Forum* 27(2): 95–113.

Myers, Fred R. and Donald Brenneis. 1991. Introduction: Language and politics in the Pacific. In *Dangerous words: Language and politics in the Pacific*. Donald Brenneis and Fred R. Myers, eds. Pp.1–29. New York: New York University Press.

Nadasdy, Paul. 2007. The gift in the animal: The ontology of hunting and human-animal sociality. *American Ethnologist* 34(1): 25–43.

Narokobi, Bernard. n.d. *The history of Wautogig village.* Lise M. Dobrin, ed. Online at http://www.arapesh.org/exist/cocoon/arapesh/howv/home.

Ohnuki-Tierney, Emiko. 1987. *The monkey as mirror: Symbolic transformations in Japanese history and ritual.* Princeton: Princeton University Press.

Ojoade, J. Olowo. 1990. Nigerian cultural attitudes to the dog. In *Signifying animals: Human meaning in the natural world.* R. G. Willis, ed. Pp. 215–221. London: Unwin Hyman.

Rice, Keren. 1989. *A grammar of Slave.* New York: de Gruyter Mouton.

Savishinsky, Joel. 1990. The Hare and their dogs: Human–animal bonds in an Arctic community. *The World and I* 5(3): 643–653.

Schneider, Katharina. 2013. Pigs, fish, and birds: Toward multispecies ethnography in Melanesia. *Environment and Society: Advances in Research* 4: 25–40.

Slone, Thomas H., transl. and ed. 2001. *One thousand one Papua New Guinean nights. Folktalkes from Wantok newspaper, Volume 1. Tales from 1972–1985.* Oakland, CA: Masalai Press.

Slotta, James. 2023. *Anarchy and the art of listening: The politics and pragmatics of reception in Papua New Guinea.* Ithaca: Cornell University Press.

Smith, Benjamin. 2012. Language and the frontiers of the human: Aymara animal-oriented interjections and the mediation of mind. *American Ethnologist* 39(2): 313–324.

Stewart Strathern, Pamela J. and Andrew Strathern. 2023. *Dogs and their affiliations.* University of Pittsburgh unpublished manuscript.

Tambiah, Stanley J. 1969. Animals are good to think and good to prohibit. *Ethnology* 8(4): 423–459.

Tannen, Deborah. 2004. Talking the dog: Framing pets as interactional resources in family discourse. *Research on Language and Social Interaction* 37(4): 399–420.

Tuzin, Donald. 1997. *The cassowary's revenge: The life and death of masculinity in a New Guinea society.* Chicago: University of Chicago Press.

3 Dogs in Majuro

Pests, Pets, and Perils

Jordan Prokosch

Introduction

When I researched in Majuro my interest wasn't in dogs. Nevertheless, dogs inserted themselves into my research experience and have a role to play in understanding Marshallese life. When they introduced this species, people of the past altered the atollscape of the present. This alteration opens new possibilities of action (adopting a stray), new hazards (dog attacks), new hassles (dog poop), and new life experiences, positive or negative. These new life experiences contribute to individual mental models and, if they become ubiquitous, form new cultural models (see Prokosch 2021; Holland and Quinn 1987).

Dogs are not just *Canis familiaris*; they are also a schema, a bundle of interconnected ideas. These ideas include functional, ethical, legal, and even emotional concepts. For example, one schema holds that dogs are "good boys", cute, goofy, and ultimately noble creatures who are occasionally victimized by stupid, lazy, or cruel human beings. These concepts become institutionally embedded, e.g., in laws, shelters, professions, and unwritten norms that are socially enforced. In the US, if you kick a dog to death, you may face criminal charges, and many people will treat you like a monster. If your dog is sick, you are supposed to take them to the vet and deal with the costs as part of your responsibilities as a pet owner. You are supposed to spay/neuter them unless you plan to breed them, a procedure readily available with charitable assistance if it is a financial difficulty. So then, what do dogs mean for Marshall Islanders?

Common Experiences

Pests

When I picture a Majuro dog, it's a skittish, pathetic thing with mangy fur and a beaten look, scavenging at the outer edges of stores that serve food, or around garbage, or down alleys. Under the right circumstances, these dogs could be threatening, but they were mostly just pathetic. That was

DOI: 10.4324/9781003658696-3

my first encounter with dogs in Majuro, as an ever-present, unpleasant part of the downtown cityscape. My reaction to these dogs was disgust tinged with fear, which could quickly turn to anger if pushed. I didn't want to touch them, and I didn't want them to touch me. In America, I like dogs. I like dogs with lustrous coats, wagging tails, and joyful energy. I did not like these Marshallese animals. The difference reminds me of a rat I used to own as a child. It was black and silver, well-fed, beautiful, well-mannered, and a pleasure to see and touch. Just because I owned and loved that rat doesn't mean I want to snuggle a wild one in an alley.

This personal experience of dogs resonated well with my Marshallese friends and family. The generally unpleasant relationship Marshall Islanders have with dogs is even linguistically encoded. The Marshallese word for dog is *kidu*, an onomatopoeia reminiscent of a low threatening growl. The dogs are smelly, dirty, and annoying, like the litter, the flooding, or the flies. Flies are a good point of comparison. Islanders express disgust at the common swarms of midday flies and worry that they will ruin the food, but their solution is to cover the food when not eating and, when eating, to expend minimal effort fanning them away. Similarly, stray dogs are corralled from important areas but otherwise ignored, if they bother someone, they are driven off temporarily with thrown stones. If they're commented on at all, it's with annoyance or embarrassment.

Children sometimes take a more active interest in the animals, though not an affectionate one. I've seen groups of children tease dogs with food, beat them with sticks, try to hurt them with stones, and hold puppies underwater. These acts of violence are met with little criticism from adults; the only exception I observed was when an adult man told his younger relatives to stop taunting a street puppy they'd found; he turned to me right after and explained in English, "They are teaching it to be mean." I don't think he would have intervened if I wasn't there, and his explanation had little to do with the animal's suffering. He thought it was wrong because of the long-term consequences for the community.

When it came to these animals, empathy was in short supply. In Lohmann's (2011) terms, they do not extend empathetic imagination to street dogs. This tends to be the case for all animals. Vegetarianism based on preventing animal suffering is *ri-pālle* (American/white person) nonsense. Animals are not people; they don't have souls; God put them here to be useful to people. The rigid separation between person and animal is linguistically marked. All animals are categorized as *men in mour*, living things. People are *armij*, a unique term. I was once forcefully corrected when I called myself a living thing in Marshallese. The problem with my statement wasn't with *mour* (life), something that people and animals share; it was with *men* (thing). People are not, and can never be, things. Conversely, things, even living things, are not people. Dogs readily give strong emotional signals that humans easily perceive (Kujala 2017). Nevertheless, if a person fails to engage in the empathetic process and will not entertain that the dog's suffering is real in the way a person's is, then the further extension of sympathy is impossible (Fuchs 2019).

The local expressions of the pan-pacific concept of love-compassion-concern-pity (Hollan and Throop 2011), are *iokwe e* (hello/love/rainbow to it) or *buromoj e* (sad/longing for it). I rarely heard these expressions concerning animals, even when they suffered greatly. When Marshallese catch turtles for a feast, they leave them to slowly die on their backs to better preserve the meat. I was told a story about an American school teacher who became enraged at this display of animal suffering and freed the turtles. The village had no sympathy for her. She'd ruined a party, disrespected her hosts, and personally insulted the chief, all for some beasts. She was fired soon after. The storyteller was using this tale to warn me about being self-righteous as an American in the Marshall Islands. I understood the message and agreed that her behavior was unacceptable, but I empathized privately. She felt for those sea turtles and ultimately sacrificed her reputation and job to alleviate their suffering. If we were talking about suffering people, her heroism would be self-evident. For most Marshallese, that comparison is baffling, even insulting. Turtles are more like mosquitos than people. While there are many other valid reasons for the disproportionate participation of Marshallese in the US meat industry (see Duke 2017), I suspect this lack of empathetic extension to animals is advantageous when working in a slaughterhouse.

I don't want to over-generalize with the claim that Marshallese never extend empathy to animals. I think animals are typically, but not universally, objectified and placed in a category of being undeserving of empathy and the moral obligations extended to fellow humans. I once visited the Woodland Park Zoo with my wife, who is Marshallese. The gorilla exhibit showed a video of a gorilla mother nursing her newborn. She had the same posture and used the same rocking gesture my wife did when she nursed. Across the glass barrier, the live gorilla was curled up and hugging herself, which I interpreted as a response to the cold weather. She read it as shyness because a crowd of strange people were staring at her. These two human-like gestures triggered the imaginative process of anthropomorphization and, with it, the extension of her empathetic imagination (Airenti 2015). It was not an animal anymore, not a living *thing*, it was a kind of person.

With that shift came a moral obligation to act. She insisted that it was wrong to cage gorillas, just like it would be wrong to cage people who didn't do anything wrong. She felt *buromoj* towards them and was disturbed for some time after. She still feels uncomfortable when she's reminded of caged gorillas.

Empathy isn't just feeling, at least not in the Marshall Islands; to be genuine, it must be expressed through action, primarily caregiving and feeding (see Hollan and Throop 2011). People deserve empathy and must be cared for. I routinely hear Marshall Islanders proudly declare that they have no starving or homeless people on the islands. There are powerful cultural prohibitions against hoarding food. To feel empathy for stray dogs would mean accepting them as people and taking responsibility for them. It would mean taking precious food from your own plate for their benefit. For the poorest Marshallese, who spend the most time with stray animals but can also least afford to care for them, the failure to extend empathy towards them may be a practical necessity. For those who can afford a pet, the preference is not to adopt an animal in need of care, but to import new pure-bred dogs from America.

Pets and Guardians

A few days after I arrived in Majuro for the first time, I met a very different kind of dog. I was invited to a picnic party by a wealthy Marshallese man. He owned a pit bull, newly imported from Hawaii. It looked healthy and very strong. He laughed about how he liked having the biggest house and biggest truck, and now he has the biggest dog. He brought it to the picnic and encouraged me to approach the dog with him so it would know my scent. As soon as I came close, it pulled at its chain and snapped its jaws at me. His wife made a displeased remark about how dangerous the animal was, but he laughed it off. That was the closest that dog ever came to attacking me. Its owner never trained the dog not to attack people, but he kept it caged or chained whenever I saw it. Other dog owners in Majuro were less considerate. I soon learned that strays in Majuro were an irritant, but pets were the real danger.

I did not research the issue systematically, but pet bites seemed very common on the island. I heard many stories, mostly told in a comedic style, of dog attacks. Seemingly, everyone has a pantomime of startled jumps, dragged legs, and pained behinds. These jokes reflect a darker reality. My stepdaughter struggled for years with a paralyzing phobia of dogs. Generally, the strays are non-territorial and lack the strength to menace a person. It's the pets, large imported breeds with the strength and energy that come from a good diet, that would protectively shoot out of open doorways, around fences, and down alleys to attack pedestrians.

Pets in Majuro are largely the domain of the upper class. Class in the capital combines multiple spheres of power: traditional lineage, religious authority, political power, education, and monetary wealth. These spheres interlock so that a member of a chiefly lineage can use their land rights to make wealth through renting, spend that wealth on tutors, give generously to the church, and then fund a campaign that leans heavily on their culturally mandated leadership role. These upper-class Marshallese tend to be more Americanized than the lower class *ri-jerbal* (literally workers). By "Americanized" I mean they have more experiences from and of America. They are online more often, learn at an American-model English language private school, speak fluent English, watch more American movies, often have close American relatives, and have spent time in America. It's expected that they'll have a few *ri-pālle* quirks, such as keeping animals as *nājin,* a term that initially confused me because it applied to someone's children, dolls, electronics, and pets. The term indicates that these things are important to someone and must be protected and indulged for their sake. Even if you don't particularly care about it, mistreatment of anyone's *nājin* will mean serious trouble. If it's someone with power, as is often the case for pet dogs, the consequences could be disastrous. These concerns were a more prominent deterrent to animal cruelty than any thinking about the animal's internal experience in itself.

Upper-class Marshallese might be more Americanized, but they are not Americans. In the United States, enthusiastic displays of affection, such as shouting joyfully, petting, and roughhousing with a dog, are common, even for non-owners. Public displays of affection are embarrassing and thus a rarity among Marshallese. While many American pet owners allow their dogs inside the home, they are strictly outdoor animals for Marshall Islanders due to the general perception that they are inherently dirty and high cultural standards for domestic cleanliness. Americans typically walk outside with their dogs. I can only think of one Marshallese man who would walk dogs around Majuro, an act that caused general distress and contributed to his reputation for craziness. Obedience training is common in the US, but not for dog owners in the Marshall Islands. The infrastructure for dog training is non-existent; there are no dog classes, dog parks, professional dog trainers, or dog training aids for sale. None of these things are strictly necessary to train a dog, and upper-class Marshallese would have access to dog training information online and may have been exposed to schemas surrounding the training of dogs. However, the few times I discussed it with upper-class Marshallese, they were ambivalent; disciplined dogs are cool but not worth the work.

In their definition of pet animals, DeMello (2021) points to animal naming, enjoyment of the animal's presence, grief at the animal's death, and physical affection as key indicators of pet status. Dog owners in the RMI name their dogs. They rarely openly show affection towards the animals, do not openly grieve them, and do not give them funerary rights.

In light of these contrasts, "pet" may be the incorrect term for owned and well-cared-for dogs in Majuro. I believe the word still applies for a few reasons: (1) the term *nājin* implies attachment, affection, and sadness if they are lost. (2) They are being fed well, which, for Marshall Islanders, is a highly symbolic act of love and care. (3) The overall norm is limited public displays of affection, even among parents and children or between couples who are deeply in love. (4) I can recall two instances when an upper-class Marshallese person (both teenagers) expressed warm feelings about a dog, in contrast to the hundreds of negative comments I've heard from lower-class Marshallese. This is far from conclusive evidence, but it is enough for me to suspect there's a larger pattern. While I believe they provide pleasure and companionship, their primary purpose is to be working guardians.

In contrast to the US, where you can find dogs of all shapes and sizes, Marshallese pet dogs are large, aggressive breeds. This is because, in most cases, the undisciplined dogs are doing a useful job that requires size and aggression. When a pet dog shoots out of an open gate and attacks a pedestrian, it acts protectively, as intended, but just a little too zealously. In downtown Majuro, the rich and poor aren't segregated, large, beautiful homes and walled compounds jut out from a sea of slum shacks. Groups of young children wander the streets and play throughout the day, and, more menacingly, intoxicated young men replace them at night. Explicitly, the dogs are acting as guards against these unruly men, potential thieves, and rapists.

They may also protect against less acceptable targets, Berman's (2014, 2020) work points to some of the tensions and contradictions in Marshallese culture, such as the cultural expectation of ruinous generosity and the reality of people not wanting to give away things they value. I suspect (though I have not been told, it would be a shameful thing to admit) that the wealthy of Majuro would rather not have groups of poor children turning their expensively imported and laboriously cultivated yards into a playground, or hungry old women dropping by asking if they have food to spare. However, they want even less to be seen as not wanting these things, to be the kind of monster that would run from their house to shoo away the young and the needy. Like blaming children for lost food (Berman 2020), the dogs present a more acceptable excuse and a removed focus for blame, fear, and resentment. Aggressive breeds don't need to be trained at all to exhibit this behavior; they just need to be left alone. It's the dog's fault, the dumb animal. Even if the owner is to blame for negligence, this is a much more forgivable wrong than outright stingy or aggressive behavior.

There are no real negative consequences for the owners of dogs who bite people. Most often, the victim does not even tell anyone in an official capacity. They don't want to "make problems" and get a bad reputation, they fear repercussions from offending a powerful person and do not believe the owner will be punished. I agree with them; I did not report my own dog attack. Why mark myself as a troublemaker? Why conflict with someone who could get me fired or kicked off the island? I knew enough, even then, to know that a police officer would politely hear my report and act concerned, and then he would discard it.

Perils

About six months into my first time living in the city, I was attacked by a pack of healthy, strong dogs outside the home of a wealthy Marshallese family. The home had walls and a gate, but the gate was left open, and the dogs were allowed to spill onto the footpath, where they would often nap. I needed to use that path every day to get from my work to my home. If I came close, they would become aggressive, circling and barking. A man showed me how to drive them back behind the gate by throwing stones, which they would sullenly respond to. If they weren't too close to the path, I preferred to leave them where they were and skirt around as far as I could. One day, I misjudged, and the dogs attacked. I used my backpack as a shield and then ran, but I was bitten on the calf. Thankfully, I was wearing pants, and the bite on my leg wasn't a serious wound. From then on, I threw rocks when I had to cross. I also stopped walking home as much. The short taxi ride was worth the price of not dealing with the dogs again.

I have a mental map of Majuro, particularly the Delap-Uliga area, where I lived and worked. On that map, there are red spots, places that I avoid if I can, and if I can't, I go in with heightened alertness. They aren't all dog-based, but dogs were the most prominent peril. Some of these spots: (1) The place where I was bitten, (2) a narrow spot in the main road, where I must either walk very close to speeding cars or provoke a pet dog, (3) the short path between my work and the nearest store, that's often patrolled by a healthy dog, (4) a place that always floods with dirty water after rains, making it difficult to traverse. My wife is more cautious than I am, and there are places in Majuro where she will not go on foot at all. If she went to her mother's house, she'd prefer to wait 30 minutes for a taxi to take her a few thousand feet rather than walk and be harassed by the dogs that block the path.

Vehicular travel is a class signifier in Majuro. The rich own vehicles. People with a steady income use taxis. The poorest walk. The atollscape is different for each group. A driver has to worry about parking and damage to their vehicle, but otherwise, they have almost total freedom of movement, and the vehicle lets them traverse perilous spots with relative ease. They can drive past the dogs guarding upper-class homes and be greeted by the owners, a sign to the guard dogs that they should be friendly. Taxis must be hailed and charge a flat rate for each trip, adding a new wrinkle. What if you must traverse a hazard to get to the main road? Is safely traversing a bad spot worth the cost when it's a short distance? Should you pay for a taxi just to get past the dogs at the gate? Walkers are forced to take risks when no safe foot route exists. Strategies vary by individual, but there are gender and age-based patterns. Kids often develop unorthodox routes, like climbing over fences, walking on concrete walls, or cutting through private areas. Men tend to be more directly confrontational with the dogs, throwing stones and shouting to drive them back. Women will travel in groups for protection.

These decisions are not entirely about the dogs. Mode of travel is a status marker, and the city paths are dirty, leading to choices to drive or taxi very short distances even without a dog to bypass. However, I believe they impact specific navigation decisions and, over the long term, reinforce vehicles as status markers since having to throw rocks at or run away from an aggressive dog would lead to a loss of dignity. Dogs came up frequently in conversation as a good reason to flag down a taxi or head back to a car instead of walking down a particular street. There may be observer bias, the dogs serving as a great excuse to give the *ri-pālle* with a tin-ear for status considerations. On the other hand, children, who are already low status and will happily play in the dirt paths for hours, were highly aware of the location of all the dogs in their neighborhood and their relative levels of aggression. They would warn me away from specific spots and tell me to either go a different route or return to the main road and get a taxi.

They are the experts. Lower-class children, who play outside unsupervised much more often than their upper-class counterparts, are highly vulnerable to dog attacks. Physically, they are smaller, weaker, slower, less intimidating, and socially, their voices carry very little weight. Children become adults, and their childhood experiences tend to be formative, particularly experiences that involve extreme stress (Henckens et al. 2009). As they grow into young adults, these children may have internalized the fear and impotent rage that came with encountering packs of hostile animals that, at the time, approached their own size.

The lack of empathy, and sometimes hostility, towards strays makes more sense in this light. Pet dogs are large, strong, aggressive, and protected by the social status of their owners. Stray dogs are thin and weak, mangy, unloved, and non-threatening. They are both dogs, however, and when most significant experiences of dogs in the culture, particularly among children, are negative and potentially traumatic, these experiences may become part of the overall schema (see Cason et al. 2002). The poor of Majuro grow to hate dogs as a whole. The rich do not share this experience and likely have some positive experiences with their pets or the pets of their upper-class friends.

Uncommon Experiences

Food

In this volume (Chapter 9), Carucci briefly describes the Enewetak practice of eating dogs and mentions that it is prohibited among most Marshall Islanders. Nevertheless, I regularly heard lurid stories of dog eating, told by non-Enewetak Marshallese trying to shock, impress, and measure me as a sympathetic Marshallese-speaking American. I'm sure these stories were exaggerated, but I am also confident that many Marshall Islanders have eaten dogs. In the city, groups of young men (and the occasional woman) gather at night and engage in many prohibited behaviors, including heavy drinking, smoking, huffing chemicals, and using other intoxicants when available (see Carucci 2019). It was these groups, I've been told, that kill dogs at night and roast them over a fire. If that is the case (I've never seen it myself), I would characterize it as yet another act of rebellious camaraderie. They eat the dogs together because dog eating is prohibited.

I also think it's an extension of the harassment children inflict on the animals and revenge for the harassment pet dogs have inflicted on them. Killing and eating a dog is the ultimate assertion that it is just an animal. It also asserts, among people who have almost certainly avoided and run from dogs many times in the past, that they are strong, fearless, and capable, Marshallese traits of masculinity that they dramatically embrace as they transition from youth to men (Dvorak 2008). I also wonder if dogs being associated with America plays a role in both the prohibition against eating dogs and the violation of that prohibition. In Majuro, the feelings towards America are strong and ambivalent, full of admiration, affection, inspiration, fear, disdain, and defiance (Prokosch 2021). Maybe eating dogs lets young men rebel against the distant specter of American judgment alongside the closer criticism of their elders.

Civic Problems

Most islanders dislike the dogs on Majuro. The strays are ugly, annoying, dirty, and embarrassing. The pets are a hazard to pedestrians, particularly young children. They're a problem in the city, and since people's experiences of them are overwhelmingly negative and there is little empathy towards them, it seems like there should be a solution. Why are they allowed to live on the island? The answer lies, I believe, in several factors: class dynamics, the Marshallese model of America, limitations of the Marshallese government, and the indifferent relationship the majority of Marshall Islanders have with their capital.

The dogs harm the masses but benefit the upper class. Any action against the most dangerous dogs threatens the *nājin* of someone important. The local and national government is dominated by the wealthy landowners of Majuro, their relatives, and their allies, the same group that uses guard dogs. The strays are unprotected; they even occasionally bother upper-class Marshallese. I also do not believe it would be difficult to find islanders willing to kill them. If a dog did manage to bite a Marshall Islander with significant political power, I can even imagine that the strays could mysteriously disappear like the dogs in Āpia (Feinberg and Pyrek, this volume Chapter 6). However, any solution that does not deal with the pets would do little to reduce the number of dog attacks, and the pets would eventually produce the strays again. There would also be opposition to a culling based on international perception.

Marshallese of all classes are leery of openly hurting America's *nājin*, the animal the most powerful nation in the world seems to irrationally favor over all others. As former colonial subjects, Marshall Islanders tend to be aware of how things work in the United States, and of the opinions of a typical American, but this awareness has systematic biases. So, in my decade of experience living with Marshall Islanders, I've never met one who isn't aware that *ri-pālle,* hilariously treat their dogs like they're people and, more frighteningly, will judge and even try to punish you if you do not. It's less well-known that Americans also employ animal control methods, which include catching and sometimes killing stray dogs (see Morrison, this volume Chapter 4). "Americans love dogs" may be a stereotype, but if so, it's a stereotype with power for many Marshallese, who are sensitive about negative judgments.

What about capturing the dogs and sterilizing them? That's an expensive prospect that takes specialized knowledge and tools. They would need to employ dog catchers, import foreign experts, feed, and care for the strays. How much money is the Marshallese government willing to spend on animals when they have people to care for? Their budget is extremely limited and largely fixed (Fuchs 2019). The dogs are an annoyance, but they are low on the list of government priorities.

Could volunteers handle the dog problem? Church youth organizations are a possibility; sometimes they do volunteer work to improve the city. However, for all the litter cleaning campaigns they do, the streets are still dirty. They are outnumbered by disinterested islanders. Communal controls in the outer islands tend to fall apart in Majuro. Too many neighbors are strangers, and too few people feel a strong connection to the land. Most islanders that I interviewed on Majuro had predominantly negative things to say about the city (Prokosch 2021). There are Majuro islanders on Majuro, but most people are from elsewhere. The city, for them, is transitional. A place to stay a while, go to school, make some money, and then either return to their real home or continue to America. They generally aren't interested in long-term improvement projects. The dogs reinforce this sense that the city is something to be temporarily endured.

Thus, while the dogs in Majuro are broadly recognized as a problem, without some greater social movement to change things, it will continue as it is. Like a fly buzzing around their food, it gets an annoyed wave, and nothing more.

Reflections

As Lohmann writes in Chapter 8 of this volume, dogs are highly responsive to the human cultures around them. The stray dogs adapt to a semi-hostile human environment; they develop scavenging strategies, become reactive to rocks and shouts, and learn how to hide and eke out a meager existence at the edges of the city.

If Marshallese thought about them differently, either with more care or more systematic hostility, they would not exist in such prominent numbers. The pet dogs have less to learn; they bond with their masters and follow the protective instincts they were born with, for which they are richly rewarded in food and safety. They also only live as they do because of the culture they live in. They would not be so common if the rich segregated themselves more from the poor or if there were other ways to deter intruders. They would be more controlled if their owners felt more social pressure to control them.

The culture, in turn, responds to them. The way Marshallese think of dogs is influenced by how they are present in their lives as pests, pets, and perils. Dogs play a prominent role in mental mapping and navigating the city. Disgusting encounters with stray dogs and stressful encounters with pets acting as guardians become ubiquitous experiences for most residents. For the rich, the more prominent experience is of a loyal guardian, and potentially for the younger generation, as a loving pet, reinforced by the American media they consume. This gap in experience contributes to class divisions. As a species that reacts to and acts on culture, dogs can reveal much about the lives of their humans.

References

Airenti, Gabriella. 2015. The cognitive bases of anthropomorphism: From relatedness to empathy. *International Journal of Social Robotics* 7(1): 117–127.

Berman, Elise. 2014. Holding on: Adoption, kinship tensions, and pregnancy in the Marshall Islands. *American Anthropologist* 116(3): 578–590.

Berman, Elise. 2020. Avoiding sharing: How people help each other get out of giving. *Current Anthropology* 61(2): 219–239.

Carucci, Laurence Marshall. 2019. From drunken demeanour to doping: Shifting parameters of maturation among Marshall Islanders. In *Pacific youth: Local and global futures*. Helen Lee, ed. Pp. 203–218. Canberra: Australian National University Press.

Cason, Dana R., Patricia A. Resick, and Terri L. Weaver. 2002. Schematic integration of traumatic events. *Clinical Psychology Review* 22(1): 131–153.

DeMello, Margo. 2021. *Animals and society: An introduction to human–animal studies*. New York: Columbia University Press. doi:10.7312/deme19484.

Duke, Michael R. 2017. Immigrants in small cities: Opportunities and challenges. *The Hooks Policy Papers. Vol. 2017*, article 3. https://digitalcommons.memphis.edu/hookspolicypapers/vol2017/iss1/3.

Dvorak, Greg. 2008. "The Martial islands": Making Marshallese masculinities between American and Japanese militarism. *The Contemporary Pacific* 20(1): 55–86.

Fuchs, Thomas. 2019. Empathy, group identity, and the mechanisms of exclusion: An investigation into the limits of empathy. *Topoi* 38(1): 239–250.

Henckens, Marloes J. A. G., Erno J. Hermans, Zhenwei Pu, Marian Joëls, and Guillén Fernández. 2009. Stressed memories: How acute stress affects memory formation in humans. *Journal of Neuroscience* 29(32): 10111–10119.

Hollan, Douglas W. and C. Jason Throop. 2011. The anthropology of empathy: Introduction. In *The anthropology of empathy: Experiencing the lives of others in Pacific societies*. Douglas W. Hollan and C. Jason Throop, eds. Pp. 1–21. New York: Berghahn.

Holland, Dorothy, and Naomi Quinn. 1987. *Cultural models in language and thought*. Cambridge: Cambridge University Press.

Kujala, Miiamaaria V. 2017. Canine emotions as seen through human social cognition. *Animal Sentience* 2(14): 1.

Lohmann, Roger Ivar. 2011. Empathetic perception and imagination among the Asabano: Lessons for anthropology. In *The anthropology of empathy: Experiencing the lives of others in Pacific societies.* Douglas W. Hollan and C. Jason Throop, eds. Pp. 95–116. New York: Berghahn.

Prokosch, Jordan. 2021. *Marshallese post-colonial experience, preservation and innovation in contemporary Oceania.* Ph.D. thesis, Washington State University, Pullman.

4 More Aloha for the Dogs

The Cultural Context of Dogs in Hawaii

Lynn Morrison

Introduction

This chapter on the human–canine interactions, or the "culture of dogs" in Hawaii, represents a melding of my research experience in human health and my personal love of dogs. As a Canadian physical and biomedical anthropologist, my Border Collie/Husky Kokanee and I moved to Hilo, Hawai'i 24 years ago to teach and conduct research at the University of Hawai'i at Hilo. The opportunity to continue my work on women with HIV/AIDS and to start new research on the evolutionary significance of menstruation and cross-cultural differences in the experience of menopause was exciting and fulfilling. However, outside of work, I was dismayed at the absence of dog parks and the prohibition of dogs from hiking trails and local beaches. I came from a dog-loving culture where cafés put out dog bowls for our canine family members and dog parks were abundant. I struggled deeply with this new culture I found myself in where my dog was not welcome. In the mornings, Kokanee and I ran in our neighborhood hoping to find dog friends but what we encountered were large pit bulls, Rottweilers, and a variety of hound dogs tied on short chains. I adjusted my running route for safety purposes and to protect my heart from starting the day with so much despondency. I volunteered at the local shelter where few rules existed as there were few volunteers at that time. Kokanee and I eventually found a dog friend on our running route but he was homeless so we adopted him. On my weekly two-and-a-half-hour commute to the west side of the island for research purposes (unrelated to dogs), I found a starving puppy infested with fleas, ticks, and worms. I adopted him, too. I came back to my lab with urine and blood samples to process for our research and, at least once a month, I also had a dog I inevitably found on the road. I fostered the dogs, brought them back to health, and found them homes. I volunteered at my young son's school where he, one of our dogs, and I conducted interactive workshops on "How to Care for Your *Ilio* [dog]". I had to take deep breaths when one child described how his uncle was really good at cutting the ears of all the pit bulls in his extended family. One of my current dogs was found

DOI: 10.4324/9781003658696-4

outside a drug house where his tail had been docked so I knew people physically modified their dog's appearances outside of veterinary care.

I continued like this for over ten years, conducting human health research for work and saving dogs on the side. More troubling was my struggle with what I perceived as a very different culture of dogs, or what Lohmann (this volume, Chapter 1) refers to as human–canine interactions, in which the values, beliefs, and lifestyles surrounding dogs was vastly different than my own and detrimentally affected dogs. A significant part of our training as anthropologists is to respect other cultures and try to understand the context and relevance of customary practices. I had travelled to many parts of the world where dogs occupied liminal spaces between object and companion. I was now living in a culture where dogs seemed to be very badly treated, and it challenged my professional training to remain objective and non-judgmental, my sense of ethics steeped in my own Canadian culture, and my personal vocation as an animal advocate. To gain a better understanding of how cultures can differ so significantly in our relationship with dogs, I began including material on the evolution of humans and dogs and the benefits of having dogs for my human evolution, ecological, and medical anthropology classes. I also supervised student internships at the local shelter. Finally, in 2013 I made the bold decision to re-orient 25 years of research on women's health to focus on methodically and systematically examining my nascent but growing idea of the "culture of dogs" in Hawaii to increase our understanding of how cultural attitudes toward dogs can differ so dramatically and how those differences impact dog welfare. My goal was to provide data to influence policymakers and animal agencies to improve dog welfare on the island. Additionally, I established Action4Animals Hawaii (2021) an animal rescue non-profit with several individuals. We work closely with animal control and shelters, sanctuaries, and spay-neuter agencies to ameliorate the life of dogs and cats on the island of Hawaii. This chapter is based on research conducted with numerous student research assistants and on years of personal observation.

As a physical anthropologist fascinated by human evolution, looking to the past was my first step in understanding how the human species evolved to have such different lifestyles with dogs. The evolutionary intersection between humans and dogs began at least 15,000 years ago and possibly even earlier based on the appearance of the first proto-dog dated to 40,000 years ago (Axelsson et al. 2013; Losey et al. 2013; Morell 2015; Shipman 2021). The latter date is controversial, with some disagreement as to when the skeletal changes are significant enough for the dog to be fully identified as a dog, or a recognizable Paleolithic dog, and not wolf-dog or dog-wolf (Shipman 2021). *Homo sapiens sapiens* evolved approximately 150,000 years ago indicating that for the latter one quarter of our existence, we have lived side-by-side, and have inter-mingled to varying degrees, with canines. Early sites of humans and dogs indicate our incipient and variable attitude toward dogs (Losey et al. 2013, Shipman 2021). For example, early burial sites from 12,000 to 14,000 years ago in Germany and Israel reveal that dogs were well-cared for with one such burial revealing a human curled around the dog (Morell 2015; Shipman 2021). We do not know if the human and canine died together or if the dog was killed to be buried with the human, but it does indicate intent to show a close relationship between the two. Archaeological evidence of the Upper Paleolithic reveals an intensification of dog–human interaction as dogs were used to hunt small animals. By the onset of agriculture in the Middle East approximately 10,000 to 12,000 years ago, the status of dogs plummeted when they were relied upon as a source of food rather than for their assistance in obtaining food (Morell 2015). As agriculturalists owned property, material items, and animals, it is likely dogs also served to guard their possessions (Lohmann, personal communication, 2023).

Using genetic and archaeological evidence Greig (this volume, Chapter 5) traces the arrival of dogs in Oceania to 3,500 years ago and identifies Papua New Guinea as a "hotspot" of subsequent dispersal of dogs of various sizes and breeds for hunting and as a food source. Polynesian voyagers brought the now domestic dog along with pigs, to Hawaii Island in pre-colonial times (Cramb 2022). Dogs can be identified in pre-historical petroglyphs and in later historical accounts by missionaries who refer to roasted dogs being a delicacy cherished by the *ali'i* (chiefs) while baked dogs, especially puppies, were enjoyed by Hawaiians in general (Titcomb 1969). Once established in Hawaii, the short-legged, pointed-eared dog was fed breadfruit and poi, made from taro, and was the preferred food source over that of pig and chicken (Luomala 1960; Titcomb 1969). Unlike other islands and atolls of Oceania where extirpation of dogs occurred, dogs persisted in Hawaii (Cramb 2022; Greig, this volume, Chapter 5). Similarly, dogs were and continue to be used as a food source in the Marshall Islands (Prokosch, this volume, Chapter 3) and specifically in the Enewetak and Ujelang atolls (see Carucci, this volume, Chapter 9).

The cultural tension between Hawaiians and Europeans about dogs dates back several hundred years when Hawaiians could not understand why Europeans approved the eating of hogs but not dogs; and why one species was favored over another (Titcomb 1969). Interestingly, this speciesism debate has gained full fervor among contemporary animal activists. Currently, the island of Hawai'i exemplifies varying attitudes toward dogs with some regarded as living, sentient beings deserving of care, respect, and love, while other dogs are treated like possessions valued only to the extent they are useful, and are often poorly treated, neglected, abused, and discarded (Gray and Young 2011; Kim et al. 2018; Morrison et al. 2021, 2023; Carucci, this volume, Chapter 9). The different perspectives on dogs are based on an intersection of geographic region, culture, economics, and personal ethics and ideology. These variables influence whether dogs become household companions, guard, or hunting dogs.

According to the Animal Legal Defense Fund (ALDF), Hawai'i is in the bottom tier for animal protection laws ranking 39th out of the 50 states (ALDF 2021). Spaying and neutering is not a social norm, resulting in a profusion of unwanted dogs, especially hunting dogs (Morrison et al. 2021). Hunting dogs are sometimes starved before the hunt to improve their performance. However, if they underperform, they are readily discarded and replaced (Morrison et al. 2021:6). Hunting dogs suffering from pig gorings or other injuries are either treated in the rainforest, abandoned, or killed. Some hunters have a higher standard of care and safety for their canine hunting companions during transport and the hunt but they are in the minority. Guard dogs who are typically staked on short chains or ropes around the property are often neglected and under-stimulated. They have little shade from the sun or protection from the persistent daily rains. To be a dog lover or animal ally on the island of Hawai'i is very difficult as the neglect and overt abuse is ubiquitously tolerated by the larger community as a cultural norm. In-coming out-of-state university students and other newcomers from the continental U.S. often comment on the lack of care for animals while those from the islands become defensive about being judged by outsiders for their cultural practices. Studies have revealed that those who care for and feel protective of animals are deeply disturbed by exposure to animal cruelty (Blouin 2013; Levin 2017).

Several no-kill non-profit rescues receive abandoned dogs but cannot intake the plethora of dogs on the island who ultimately end up in two primary county animal control shelters that are persistently over capacity. With few options and little county funding, the most pragmatic and expedient outcome is euthanasia for the domestic animals entering the county shelters, primarily dogs and cats, but also other pets, like rabbits and guinea pigs, and farm animals like sheep, goats, and horses. In 2023 Hawai'i island had the second highest intake of dogs after O'ahu island, with 1857 and 6389 dogs, respectively. To note, Hawaii Island has only 1/5 the population of O'ahu but almost 1/3 of the dog intakes (World Population Review 2024). More significantly, the euthanasia rate on the island of Hawai'i is 42%, while that on O'ahu is 21 % (Best Friends 2024).

High euthanasia rates result in significant emotional and physical stress for the shelter workers who simultaneously take care of the very animals they often must euthanize (Arluke 2002; Rogelberg et al. 2007; Baran et al. 2009; Anderson et al. 2013; Andrukonis and Protopopova 2020). This has been identified as the "caring-killing paradox" and occurs among shelter workers who chose their jobs because they love animals and want to care for them but are also responsible for euthanizing them (Arluke 2002). On the island of Hawai'i, there were two main shelters at the time of the study on opposite sides of the island; the east side Kea'au shelter housed approximately 40 dogs and was in a socioeconomically challenged area, while the west side Kona shelter housed approximately 20 dogs and was in an affluent tourist destination. The two locations differ culturally with the Kea'au side having more of a local culture that includes a deeply rooted hunting and guard dog tradition while the Kona side attracts many prosperous US mainlanders who often re-locate bringing with them their companion dogs along with more positive cultural attitudes toward dogs. The Kona shelter had less guard and hunting dogs than did the Kea'au shelter. Due to the higher intake rate at Kea'au it invariably had a higher euthanasia rate (Morrison 2023). Tension arises between community members and the shelters based on the common misperception that it is the shelters' fault that so many animals are euthanized, rather than understanding that it is, in fact, a community problem (Balcom 2000).

In addition to the stress experienced by the staff, tremendous stress is felt by the dogs in the shelters, which studies have shown can be quantified biologically (Hennessy et al. 2001; Bryan et al. 2013; Heimbürge, Kanitz, and Otten 2019; Morrison 2023). Cortisol, a steroid hormone released into the bloodstream as a function of stress, has long been used by anthropologists and other researchers to quantify stress. Cortisol can be found in bodily fluids and hair. When extracted from hair, cortisol is a valid and reliable indicator of chronic stress in humans and animals (Hennesy et al. 2001; Bryan et al. 2013; Ouschan, Kuchar, and Möstl 2013; Wright, Hickman, and Laudenslager 2015; Brown 2016; Heimbürge, Kanitz, and Otten 2019; Lamon et al., 2021). Elevated cortisol levels can potentially lead to negative mental and physical health in humans and is an objective marker of stress in dogs (Beerda, Schilder, Janssen, and Mol 1996).

This chapter highlights the life of dogs and the cultural attitudes toward dogs on Hawai'i island by presenting two case studies, each conducted with four to six student research assistants. The first case study compares the attitudes toward dogs between two groups of people; those who have companion dogs and those who have guard/hunting dogs. We also explore the impediments to ameliorating the situation for the dogs, which are mired in cultural tensions and lack of political and legal support (Morrison et al. 2021). This case study relies primarily on interviews and participant-observation that I conducted with my students (Bernard 2017). The second case study examines (1) how dog culture differs between the socioeconomically different east and west sides of the island and how it affects the welfare of the dogs, and (2) analyzes cortisol, the biological indicator of stress, in the shelter employees who must simultaneously care for and euthanize dogs (Morrison 2023). Critical to this assessment of dog welfare is the comparison of the stress levels of the dogs on the east and west sides of the island, which is also an indicator of differing attitudes toward dogs. To note, the two sides of the island are not strictly delineated and each side has pockets of poverty that contribute to an overall 15.1% of people living below the poverty line (Hawaii State Office of Planning 2013). However, the east side has an overrepresentation of poverty and more drug arrests than the west side (Hawaii State Office of Planning 2013; Hawaii Police Department 2024). Concentrated affluence is more evident in Kona, on the west side, than any neighborhood on the east side due to its reputation as an up-scale tourist and retiree destination that draws on the continental population.

In Hawaiʻi, "aloha" is a warm, friendly greeting that means more than just "hello" or "goodbye" when encountering someone or parting from them. To give aloha is to give kindness, caring, compassion, and sympathy and is a pervasive ideology enacted in everyday life in the Aloha State (Pukui and Elbert 1986). As shown in this work, companion dogs receive aloha as they are given compassion and care. However, for those dogs that do not receive aloha, it is the goal of these studies to elucidate the underpinnings of the neglect of those dogs and to find ways to extend some aloha to them.

Methods

Data collection in both studies included qualitative semi-structured interviews with cortisol analysis of the shelter workers and shelter dogs in the second case study (Bryan et al. 2013; Ouschan, Kuchar, and Mostl 2013; Bernard 2017). These studies were conducted in 2014–2016 and received Institutional Review Board (Human Subjects Research) and Institutional Animal Care and Use Committee (IACUC) approval. Human participants received a $30 gift card and dog participants received dog treats.

Participants and Procedures

Case Study 1 Companion and Guard/Hunting Dogs. Recruitment for participants was on the east side of Hawaii Island only. Flyers in community centers, hunting stores and stations, veterinarian offices, and dog-walking areas as well as word-of-mouth and door-to-door calls were used to enroll people who had either companion dogs or guard/hunting dogs. Twenty (eight women and 12 men) companion dog keepers and eight (five women and three men) guard/hunting keepers participated. The companion dog keepers had free-ranging dogs with an indoor/outdoor life that involved activities with their keepers. The guard/hunting dog keepers had outdoor dogs typically chained to a doghouse or kept in a kennel. Some hunting dogs went hunting several times per year while others went more frequently. When not hunting, they were kept as guard dogs. Seventy percent of the companion dog keepers were Euro-American while fifty percent of the guard/hunting dog keepers were Hawaiian/Hawaiian mixed. The companion dog keepers had one to three dogs each while the guard/hunting dogs had three to eleven dogs each. There was a preponderance of pit bull/mastiff mixes among the guard/hunting dogs. Interview topics included describing their life with their dog(s), activities with their dogs, type of exercise (including hunting) the dog received, how they felt about their dog (s), their observations about local dogs, what they considered mistreatment of dogs, how much time dogs are tied up and length of chain or rope, and how the life of dogs could be improved (see Morrison et al. 2021 for full details on methods and expanded results).

Case Study 2 Shelter Staff, Dogs, Euthanasia, and Stress. Sixteen staff were interviewed, eight (five women and three men) at the Kea'au shelter on the east side and eight (four women and four men) at the Kona shelter on the west side of Hawai'i Island. At both sites the participants included managers, front desk staff, and kennel hands. Each site had three people who were Euro-American, two of Hawaiian/Hawaiian mixed ethnicity, and three who described their ethnicity as "other". In these recorded interviews, staff were asked why they chose to work at the shelter, what their job duties involved, whether they euthanized animals, how they felt about the euthanasia and the underlying causes of it, and whether they felt the cultural and socioeconomic differences between the two sides of the island affected how dogs were treated.

To pursue our hypothesis that we would see differences in stress levels in both the shelter workers and the dogs between the two sides of the island, we obtained hair samples from the humans and dogs to extract cortisol. Using hair allows us to examine stress levels in the last three months (Bryan et al. 2013). For the dogs, this would be stress experienced before coming in to the shelter as we chose dogs that had been at the shelter less than two weeks. Twenty-two dogs from each site were assessed for a total of 44 dog samples. For a full description of the cortisol procedure see Morrison (2023).

In both case studies, the participants' names were changed to maintain confidentiality. The interviews were transcribed, coded, and analyzed using NVivo, a qualitative data analysis software (Taylor, Bogdan, and DeVault 2015; Bernard 2017; QSR NVivo 2018). After numerous iterative loops of coding and sub-coding to find commonalities within and across the interviews, this inductive analysis led to the establishment of themes. In these studies, a grounded theory approach was used that allows for common patterns to emerge from the data into themes (Strauss and Corbin 1997). Themes present qualitative data in an organized manner to explain an identified phenomenon (Mishra and Dey 2022). Interview data was compared with, and substantiated, by many hours of participant-observation in both case studies. In the first case, that included detailed notes on households, human–dog interactions, length of tie-downs, and the dogs' demeanor when approached by their keeper. In the second case study, observations included shelter workers' interactions with the dogs, with each other, and the dogs' demeanor, behavior, and overall physical appearance and state of health.

Results

This chapter presents the partial results of each case study to offer a comprehensive view of dogs' lives on the island. For more details of each study see Morrison et al. (2021; 2023).

Case Study 1

This study compared the attitudes toward dogs between companion dog-keepers and guard/hunting dog keepers to understand the context of dogs' lives and the cultural tensions that impede advocacy for the dogs. Many companion dog keepers were originally from the continental U.S. Their tangible distress as they share their observations of neighborhood dogs is the first theme. The second theme underpins the importance of the hunting tradition. The third theme highlights the socio-cultural struggle to advocate for dogs in a context lacking legal or political support for dogs in distress.

Da Dog No Can Talk: Neighborhood Dogs

Companion dog keepers felt distressed by what they observed in their neighborhoods which included seeing dogs on short chains without any respite, or others with collars growing into their necks. In some cases, participants were devastated by seeing and hearing dogs getting beaten. Josie, who has several companion dogs, stated, "We go to church … and there is a dog we pass and we look for him every Sunday and he's on, I wanna say a five-foot chain to a tree. Don't matter it's raining or whatever the weather, he's on five feet."

Dogs not roped to their dog houses are kenneled in small cages. Susie remarks on an acquaintance's dogs "they just had dogs in cages, and cages, and cages … this shouldn't be alright." As an outspoken Californian, she was dismayed at the lack of advocacy for the dogs.

Rob, who worked in a veterinary clinic, recounted how his dog Pansy came into his life:

> The Humane Society brought her in on death's door. She had been abandoned down in the subdivision just tied to an abandoned house and left for dead. They would throw food at her once a week if she was lucky … She was just a little rack of bones, skin and bones.

Households may have up to five or six dogs staked around their property to protect what they own. The less people have the more their possessions are valued as are the purebred pit bulls that protect the material items. Kanani has multiple guard dogs chained outside on very short chains because there are "too many rip-offs. That's the reason why we have our dogs". In contrast, her family had four to five small house dogs that were treated very differently than the outside guard dogs. This family had a small, dilapidated house with two generations living in it. One of their guard dogs was a blue nose pit bull, an indicator of the social and cultural value placed on guard dogs and the status it confers upon the family.

Guard dogs are also placed on vacant properties to ensure squatters do not settle there. Ikaika, who has both guard and companions dogs, is angered by the treatment of dogs and stated, "[people] leave no wata (water) … da dog gets wrap around one tree and skinny. In da sun or in da rain do not have shelter … da dog no can talk." Another participant, Janet, remarked on three dogs she encountered while doing the census that were tied to dog houses and were "skin and bones". She complained to the Humane Society several times but never knew the outcome of those dogs.

Starve 'Um, Smoke 'Um: Hunting Dogs

Tension exists among hunters who have different protocols for their dogs. Ikaika, who used to hunt and feels very strongly that hunting dogs should not be mistreated, recounted his experiences in the hunting community. He said, "They starved the dog for one day so the dog can hunt the next day." Serious injuries and death are a common occurrence when pig hunting. As Ikaika described: "You come across one big pig, your dog gonna get jacked". Jackson, who is a hunter, described how hunters starve and expose their dogs to amphetamines to improve their performance in the hunt. He explained, "I see'um cut the two-liter soda bottle … they put 'um on the dog's mouth … he [the dog] just like smoke." Putting the cut end of the soda bottle over the dog's muzzle forces the dog to inhale the drugs in the smaller mouth of the bottle. Kanani described a similar practice and added that some of the pit bulls are given steroids to increase their muscle mass.

Makana, a health-care professional who grew up hunting, is often called upon by his friends to suture hunting dogs badly wounded by boars. He no longer hunts with dogs and now derives all his joy from watching his two boxers frolic and play. Injured dogs are stapled, sutured, abandoned, or killed in the rainforest. Conscientious hunters I have met in the rainforest while hiking have a strict safety protocol for their dogs, feed them regularly, and outfit them with GPS trackers. Their dogs appear healthy and highly attuned to their keepers and responsive to basic commands. They feel that those who mistreat their dogs give hunters a bad name.

Just Keep Your Mouth Shut: Advocacy and Cultural Tensions

Animal allies, especially those originally from the continental U.S., feel deeply distressed by what they observe. As newcomers, and with few animal welfare policies in place, some feel disempowered to advocate for the animals and are concerned about possible retaliation if they were to report any abuse to animal control given the friction between mainlanders and locals. For example, Joe, who has a companion dog, stated, "You hear a lot about like, you know, 'fucking haole (a Euro-American), you move here and try to tell us what to do'. Somebody told me just keep your mouth shut for the first couple years you're here."

Josie recalled seeing a dog being beaten with a hose but was advised by a neighbor to not interfere as the abuser was drunk. She called the

Humane Society but felt badly that she did not immediately go to help the dog. When participants were asked what they thought was needed to change the conditions of many island dogs, Diane responded, "Part of it is lack of education … a generational thing where one generation teaches the next that dogs are not valuable." Others added that public service announcements informing people how to care for dogs were necessary to change the attitude toward dogs.

Case Study 2

This study explored the occupational stress of shelter workers and stress in dogs in each of two shelters, one of which (Kea'au) had significantly higher euthanasia than the other (Kona). We hypothesized that stress levels in shelter workers, measured by cortisol, would be higher in the Kea'au shelter than in the Kona shelter. Additionally, we explored whether there were differences in cortisol levels in dogs on either side of the island that may indicate different treatment of dogs. The Kona shelter, with a lower euthanasia rate, is on the more affluent side of the island than the Kea'au shelter (Morrison 2023). The data presented here includes approximately 55 hours of participant-observation at both shelters.

Stress in Shelter Workers and Dogs

The Kea'au shelter has a far greater intake of dogs than the Kona shelter with 3700 and 1353 dogs entering in each shelter, respectively (see Table 4.1). Consequently, the euthanasia rate at the Kea'au shelter is remarkably higher at 52.7% compared to 16.7% at the Kona shelter which equates to almost eight dogs per day being euthanized at Kea'au and less than one at Kona (Morrison 2023).

As can be seen Figure 4.1, the Kea'au shelter workers had a significantly higher mean cortisol level than the Kona shelter workers ($t=6.051$, $df = 13$, $p=0.001$) (Morrison 2023). Similarly, as shown in

Table 4.1 Intakes and Outcomes for the Kea'au and Kona Shelters

Intakes and Outcome	*Kea'au 2015*	*Kona 2016*	*Total*
Total Intakes	3700	1353	5053
Adopted	1091	837	1928
Redeemed	556	233	789
Euthanized	1951	226	2177
* Other	102	57	159
% Euthanized	**52.7**	**16.7**	**43.1**

*transferred, died, stolen, missing data

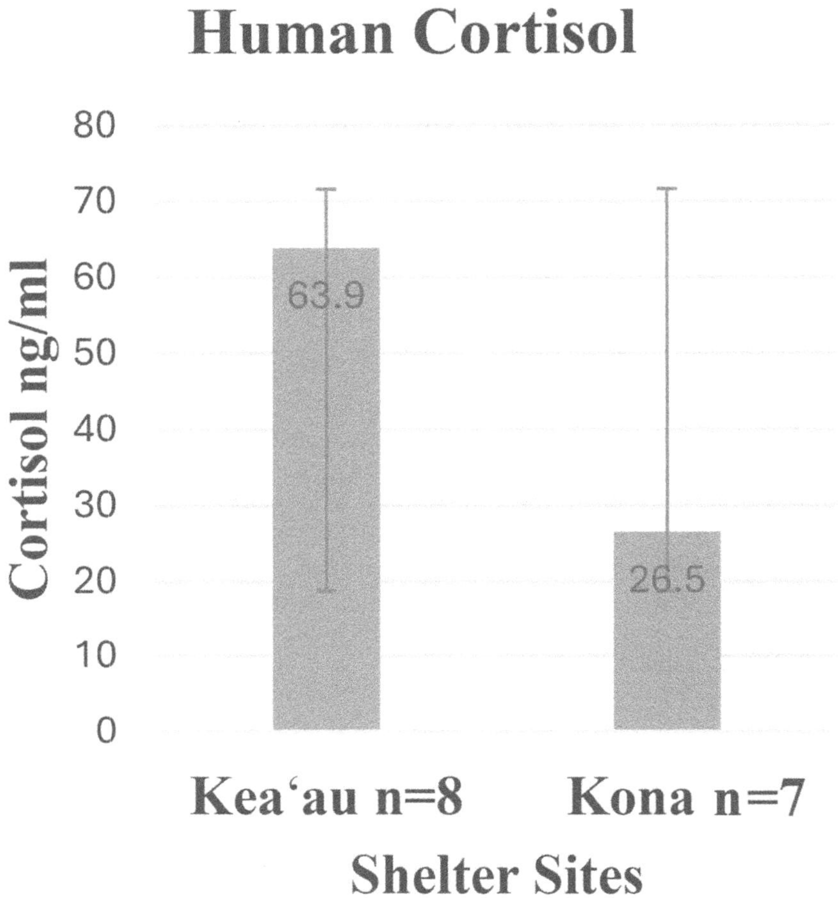

Figure 4.1 Shelter Worker Cortisol at the Kea'au and Kona Shelters

Figure 4.2, the Kea'au dogs had a significantly higher mean cortisol level than the Kona dogs (t=2.412, df=42, p =0.010) which is an indicator of chronic stress from the time they were abandoned or were living *before* entering the shelter. Based on our visual observations of their physical health and behavior, the Kona dogs were in better health compared to the Kea'au dogs, many of whom were severely undernourished with ribs showing. Although most dogs were happy to see us, a number of them were anxious and frightened of humans.

The following interview data highlights three major themes, the first of which addresses the "caring-killing paradox" (Arluke 2002). The second theme describes the shelter workers' opinions of the public's perception of shelters and shelter workers, and the third theme depicts the shelter workers' insights into the culture of dogs on Hawaii Island.

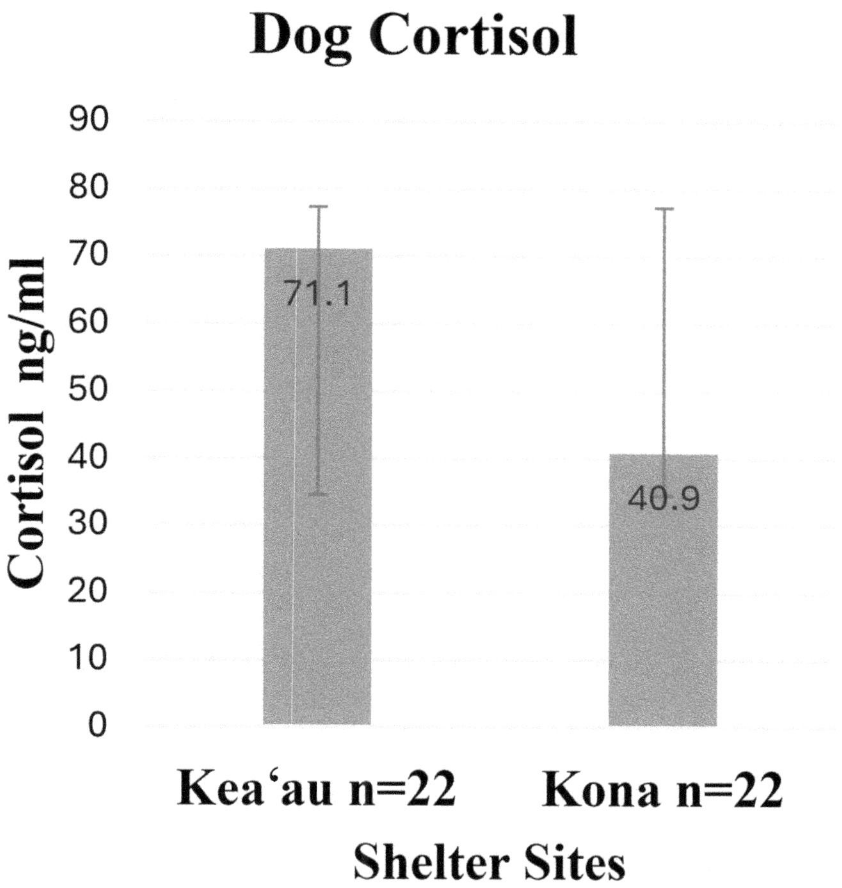

Figure 4.2 Shelter Dog Cortisol at the Kea'au and Kona Shelters

To Take Care Then to Kill: The "Caring-Killing Paradox"

Each shelter had different protocols regarding whose job responsibility it was to euthanize. In Kea'au, all staff except the front desk were mandated to euthanize whereas in Kona, only the animal control officer euthanized. However, the emotional toll of caring for particular dogs one day and then finding them absent the next day was substantial in both shelters. A Kona worker referred to the stress staff encounter as he stated, "It's actually like a pretty rough job emotionally when you see, like you get attached to a dog, and then they have to euthanize it." More poignantly, Serena, from the Kea'au shelter said, "It's been really hard, like when I first started to do the euthanasia … to take care then to kill … here you are cleaning, dropping food to them, then the next minute you're dropping them and they're gonna go in a bag."

Many shelter workers start their work day euthanizing animals. Jimmy in Kea'au said, "I can euthanize up to 20 animals in a day." He feels that by doing the euthanasia instead of other staff, he will ensure that the animals receive compassion before dying and importantly, he will decrease the animal's suffering by doing it quickly and properly. The stress also comes from having to choose which dogs have to die. For example, Hoku at Kea'au imparts, "If I have six dogs that have to move up to adoption and I have three cages, that means three need to die." She, too, echoed Jimmy's perception that she was best at giving animals compassion and had developed expertise in euthanizing quickly and painlessly. Our observations provide evidence of the care shelter staff provided to the dogs. Observed were staff cuddling and talking sweetly to the animals and it was apparent they knew the animals' personalities well. As the Kona manager emphasized, she "will exhaust every other option before I get to euthanasia".

Public Perception and Culture of Dogs: It's a Community Problem

The shelter workers felt burdened by the negative public opinion they received in the community, local newspapers and social media. As David from the Kea'au shelter opined, the "high euthanasia rate … it's a community problem … you wanna put something nasty about us in the newspaper … but please come up with a solution."

We asked the shelter worker participants why they thought there was so much animal neglect and abuse and whether there were differences between the two sides of the island. Several participants made connections to other social issues like lack of money or drug use. Ben from Kea'au said, "I think a lot of people have issues … and are not necessarily capable of taking care of themselves, let alone another dog", while Todd from Kona said that people on the east side lack education, especially concerning taking care of dogs, and have more domestic violence. David from Kea'au reflected that the east side has more drug abuse which consequently results in the need to protect one's property, believed to be most effectively done with guard dogs.

In discussing the culture of dogs on either side of the island and its link to socioeconomics, Greg from Kea'au shared that hunting dogs are kept thin so they perform better on the hunt and people then have "meat to sell to provide for their family, as well as the dog". David from Kea'au is near retirement age and of Hawaiian descent. He is familiar with the history of the island and explained that many people on the east side are descendants of sugar plantation workers who immigrated to Hawaii from other areas of the world. Their attitudes toward dogs included mistreating and beating them and relying on guard and hunting dogs to survive.

Many informants remarked that the Kona dogs were more likely to sleep inside and not be "sleeping on cement chained to a post", as Karla from Kona commented, explaining that the east side dogs came from poverty-stricken areas. Hoku from Kea'au, whose job responsibility was to pick up unwanted dogs and litters of puppies, reflected that on the east side dogs "seem more disposable" than on the west side. Finally, Oscar from Kona remarked that people on the east side "don't want people meddling in their business" so having pit bulls and mastiffs staked out on the property ensures their privacy. Others noted that the west side is wealthier and has more Golden Retrievers that are treated as family members.

Discussion

For many people, especially those from the continental United States, the aloha state of Hawaii is iconic, conveying images of swaying palm trees and hula dancers, drinks decorated with pineapples, and long beaches with endless surf. However, there are invisible layers of distress that the animals on our island are trapped in. Similar to other parts of the world where colonization has had a damaging impact, our island struggles with social and health epidemics that include drug use, obesity, intimate partner violence, and animal abuse and neglect (Edwards, Giroux, and Okamoto 2010; Guruge, Morrison, Jayasuriya-Illesinghe and Mock 2016; Uchima, Wu, Browne, and Braun 2019). How did dogs become part of the social epidemics, why did they become so objectified, and why are there such cultural differences? Historical research substantiates that early dogs were primarily a vital food source surpassing the pig and chicken in their dietary importance. In contrast, early accounts also document the care some dogs received from Indigenous Hawaiians (Luomala 1960, Titcomb 1969). More recent history indicates that sugar plantation workers in Hawaii also used dogs as a food source or for protection (Kessler 2016), a practice also suggested by our participant David. As in other geographic regions including Pacific Islander communities, dogs are another organism to be used and dominated to increase personal survival. For example, Carucci (this volume, Chapter 9) describes several Marshallese atoll communities' changing relationship with dogs over time and the geographic location they migrate to. On Ujelang and Enewetak atolls, dogs were periodically completely absent because they were eaten or when present, were mistreated and homeless. Ujelang/Enewetak people who came to the island of Hawaii adopted the local Hawaiian practice of having large pit bull guard dogs who receive little interaction or care and are sometimes overtly abused (Carucci, this volume, Chapter 9). Using dogs as a food source is not unique to the Pacific Islands, as Gray and Young (2011) illustrate in their cross-cultural examination of animals. They found that in numerous societies, dogs primarily function to guard property, assist in the hunt, or are a food source.

Several authors in this volume, myself included, make the point that perceiving dogs as companions, beasts of burden, pests, or sources of food is culturally-based. The values and beliefs associated with dogs are generationally and cross-culturally transmitted (Carucci, this volume, Chapter 9).

Lohmann (this volume, Chapter 1) introduces the idea of the human–canine interspecies culture, noting among its manifestations the social expectations and roles of both humans and dogs. The culture of dogs, the expression of people's values, beliefs, ideology, practices, and the extent of integration of dogs into everyday life, is sometimes static while at other times dynamic depending on the circumstances of place and time. Some of the cultural tension surrounding the treatment of dogs in Hawaii may also be due to an inability to understand another individual's level of empathy, expressed as concern and compassion, or lack of it (Lohmann 2011). The human–canine interspecies culture is malleable and diverse and can present differently within a community and even within a household, as exemplified by Kanani, who had small free-ranging dogs in the house while she also had three outdoor pit bulls staked to their dog houses.

The participants in Case Study 1 depicted the conditions of undernourished neighborhood dogs that were left unattended for prolonged periods of time on short chains. Other examples revealed dogs that were beaten with hoses or subjected to other forms of severe neglect. I once found a dog at a local beach park who was so weakened by starvation that she could no longer walk. The veterinarian at the Humane Society told me her insides had liquified and she was putrefying despite still being alive which explained the smell she emitted. The only option was to euthanize her immediately. As dog rescuers, it is not uncommon to find dogs whose collars have grown into their necks and need to be surgically removed. My personal observations of animal neglect and those of the participants in Case Study 1 are limited to the east side of the island. Despite this regional limitation, the findings of Case Study 1 contribute to a better understanding of the utilitarian value of dogs in the hunt to provide food for families and for the dogs. The misuse of hunting dogs by withholding food and giving them drugs so they can perform better on the hunt is also depicted.

The overall interview and cortisol data suggest a difference between the two sides of the island in human–canine interactions and attitudes toward dog, broadly encapsulated as differing cultures of dog. As stated by several shelter workers in Case Study 2, the east side of Hawaii is more vulnerable to social epidemics such as drug use and poverty than the west side. The shelter workers also felt that people on the east side struggled with a lack of education and an overall inability to take care of family members, leaving dogs in even more desperate situations. The socio-economic circumstances on the east side not only influence how dogs are treated but also influence the breeds that are sought after, such as the pit bulls and mastiffs who are relied upon to guard people's legal and illegal possessions.

Awareness of the cultural tension in Hawaii between newcomers from the continental U.S. and the local populations was a deterrent to our participants to advocate for the dogs. The findings in Case Study 1 describe how participants struggled with cultural appropriateness in speaking out, the fear of potential recriminations, and the lack of confidence to overtly support dogs in distress. Our participants perceived non-interference to be culturally valued on the island of Hawaii. Many properties, especially on the east side have as many "No Trespassing" signs as they have pit bulls tied along the perimeter of the property. Much of the local population adheres to the cultural idiom that if it does not affect you or your family, it is not your place to voice concerns. Participants, especially those that were "haoles" (Euro-Americans) and newcomers were concerned about transgressing that cultural custom despite how distressed they were about maltreated dogs. Several called the local Humane Society in the hopes that the abused dogs would receive assistance. Borrowing from Blouin's (2013) terminology, these animal allies who had companion dogs had a "humanistic" and "protectionistic" orientation, which explains why they were so deeply disturbed by observing animal cruelty, or otherwise "dominionistic" behavior, on a daily or weekly basis (Blouin 2013). Those who attribute a higher value to dogs and empathize with them suffer from a significant emotional burden in the face of abject mistreatment and neglect of dogs evinced in the comments of our community participants and shelter workers (Lohmann 2011; McConnell et al. 2011).

The number of unwanted dogs is especially high on Hawaii Island compared to the other islands in the state despite the significant efforts of animal-oriented non-profit organizations providing free spay/neuter services throughout the year. The result is close to two thousand dogs entering overburdened and underfunded shelters yearly with 43% of them getting euthanized. A common public misperception holding the shelter workers responsible for euthanasia is very inaccurate as the shelter workers were deeply distressed by the "caring-killing paradox" they experienced (Arluke 2002; Balcom 2000; Rogelberg et al. 2007; Baran et al. 2009; Anderson et al. 2013; Andrukonis and Protopopova 2020). The shelter workers were observed caring for the dogs, knew their names, personalities, and the story of how each dog entered the shelter whether it was owner surrender or if they were picked up as strays by animal control. Many community members are ignorant of the neglect of guard and hunting dogs and backyard breeding because of a very opaque cultural lens of acceptance but are quick to blame the shelters for the outcome of injured and unwanted dogs. Public protests and disdain of the shelters place an unfair responsibility upon the shelter workers for what ultimately is a social, political, and legal animal welfare issue that should be addressed by the county and the state. Notably, the dedicated animal loving shelter workers conducting the euthanasia carried a further burden of wanting to euthanize rapidly, efficiently, and with as much compassion as possible to ensure the dogs did not further suffer.

The significantly higher cortisol rates of the Kea'au shelter workers, who euthanize a disproportionate number of dogs compared to the Kona shelter workers, give evidence of the stress response likely associated with having to kill animals they care for. There are very few occupations that present an equally distressing conundrum or that involve ending life in such a purposeful and systematic manner. Both the east-side shelter workers and the dogs had significantly higher stress rates as measured by cortisol than their west-side counterparts highlighting critical differences in shelter environments as an extension of the communities within which they operate. This becomes both an animal welfare and a public health concern. The human and dog cortisol results must be regarded with caution, however, as it is difficult to eliminate other sources of stress. For example, it is unknown what stresses the shelter workers encounter in their personal lives. Although we sampled most of the staff at each shelter, a larger sample could control for stress from other areas of their life.

The focus of this chapter is on guard, hunting, and shelter dogs on Hawaii Island with specific attention on neglected and abused dogs. At the same time, many animal lovers on the island have companion dogs and those in our study reflected on how they found love, comfort, and even solace in their dogs (Morrison et al. 2021). It is unknown how many dogs exist on the island of Hawaii but a disproportionate number of them are not in healthy circumstances. Government agencies and county representatives, communities, and individuals need to be held accountable for the overabundance of animals and initiate spaying and neutering as a norm simultaneously with public pressure and culturally appropriate education on caring for dogs and animals. The sense of aloha and love of the *'aina* (the land) that influences everyday life in Hawaii does not transcend to animals. Indigenous Hawaiians and long-term locals who have acculturated have a deep reverence for the food and medicine the land and ocean offer. Materials derived from the forest traditionally used for clothing, bedding, and shelter are a significant aspect of cultural revitalization. School children learn chants to welcome the morning sun and special chants to the gods and goddesses asking permission to enter the forests. Acutely absent in this reverence for the 'aina is any mention of the domestic dog brought to the island by early voyagers. This is similar to Prokosch's (this volume, Chapter 3) description of the importance of Marshallese identity that includes love, care, and empathy but does not get conferred to dogs. These contradictions in ideology and cultural identity are perhaps easier to discern from an outsider's lens. At the crux of the matter is trying to change culturally held views about dogs so they are perceived as sentient beings, a difficult endeavor as the push to do so comes largely from the Euro-American, or "mainlander"/*haole*, community. Given the momentum and success of the cultural revitalization of language and cultural practices in Hawaii, it is an opportune time for animal welfare to become part of this significant renaissance. Culture, including revitalizations, is dynamic and can incorporate new ideas and practices with traditional ones. Teaching modules in Hawaii are incorporating concepts of "aloha" in the classroom to establish a culturally responsive pedagogy (Ebersole and Kanahele-Mossman 2020). Similarly, watchful care with aloha, as a locally and culturally relevant concept, is cautiously used in community suicide prevention (Krishnamurti 2020). These provide useful models of cultural responsiveness for public and community leaders to consider following and incorporating aloha for animals in educational programs. Physically and emotionally, dogs are dying to be recognized and valued. From a public health perspective, this would also benefit the inhabitants of Hawaii island as studies have shown that having positive relationships with dogs results in positive physical and emotional health (Beck and Katcher 2003; McNicholas et al. 2005; Schöberl et al. 2012).

Conclusion

Depicting the lives of dogs on the island of Hawai'i contributes to our understanding of cultural disparities regarding animal welfare. Dogs that become companion family dogs are a fortunate few, as significantly more become either guard dogs with little mobility or attention, hunting dogs with a high risk of injury and death, or shelter wards with a high risk of euthanasia. The two case studies presented contribute to understanding the cultural constructs of dogs on Hawai'i island, with dogs faring slightly better on the affluent west side than the more socio-economically depressed east side. Our findings reveal a "dominionistic" attitude toward dogs, especially on the east side, where dogs are regarded more as objects and property rather than sentient beings deserving care (Blouin 2013). Cultural tensions felt by newcomers from the continental U.S. inhibited them from advocating for dogs in distress. Cortisol, a biological marker of stress, provided empirical evidence of increased rates of stress in dogs on the east side. Similarly, cortisol also revealed higher rates of stress among shelter workers on the east side possibly linked to higher rates of euthanasia and public condemnation for what is a community problem shouldered by shelters.

The palpable colonial legacy on the island of Hawai'i needs to be navigated with respect, diplomacy, and deep understanding to address the root causes of socioeconomic disparities that affect animal welfare. The development of culturally aligned educational modules and public service messages using core concepts of caring for the 'aina *and* its animals and extending some aloha to the dogs could have a positive impact on both human health and dog welfare.

As Wynne (2019) passionately points out, humans have not had a longer or more significant relationship with any other animal than the dog marking the human–canine interspecies connection an especially profound evolutionary outcome. Several thousand dogs are euthanized each year on Hawaii Island and countless others are mistreated. Dogs should be recipients of some aloha that includes kindness, compassion, sympathy, and mercy. Perhaps what remains most significant in studying the culture of dogs is the myriad ways in which humans think, feel, empathize, and act toward dogs. In sharp contrast, dogs' culture regarding humans (how dogs view and respond to humans) is far more homogenous, predictable, and streamlined; dogs work for our praise, seek our compassion, and return our love. We need to learn from dogs.

Acknowledgements

Mahalo to research assistants Heather Bailey, Johanna Hill, Marina Kelley, Joey Caldwell, Stephanie Montoya, and Zaq Tman. Thanks to colleagues Julie Luis Adrian for assistance with Case Study 1 and to Dana-lynn Ko'omoa-Lange for cortisol analysis. Mahalo Becky Ostertag for statistical analysis and Kea for fur samples. Mahalo to Dr. Kale Langlas for indispensable cultural guidance and editing. Research reported in this publication was supported by the National Institute of General Medical Sciences of the National Institutes of Health under Award Number R25GM113747. The content is solely the responsibility of the author and does not necessarily represent the official views of the National Institutes of Health. The University of Hawai'i at Hilo Seed Grant also supported this research. Mahalo to Roger Lohmann for inviting me to this wonderful group of anthrodogolists! Most of the dogs in this study were euthanized. They have left their mark in this manuscript.

References

Action4Animals Hawaii. 2021. https://action4animalshawaii.com.

Anderson, Keith A., Jennifer C. Brandt, Linda K. Lord, and Elizabeth A. Miles. 2013. Euthanasia in animal shelters: Management's perspective on staff reactions and support programs. *Anthrozoös* 26(4): 569–578.

Andrukonis, Allison, and Alexandra Protopopova. 2020. Occupational health of animal shelter employees by live release rate, shelter type, and euthanasia-related decision. *Anthrozoös* 33(1): 119–131.

Animal Legal Defense Fund (ALDF). 2021. https://aldf.org/project/us-state-rankings/.

Arluke, Arnold. 2002. Managing emotions in an animal shelter. In *Animals and human society: Changing Perspectives.* 2nd edition. Aubrey Manning and James Serpell, eds. Pp. 183–203. New York: Routledge.

Axelsson, Erik, Abhirami Ratnakumar, Maja-Louise Arendt, Khurram Maqbool, Matthew T. Webster, Michele Perloski, Olof Liberg, Jon M. Arnemo, Åke Hedhammar, and Kerstin Lindblad-Toh. 2013. The genomic signature of dog domestication reveals adaptation to a starch-rich diet. *Nature* 495(7441): 360–364.

Balcom, Sarah. 2000. Legislating a solution to animal shelter euthanasia: A case study of California's controversial SB 1785. *Society & Animals* 8(1): 129–150.

Baran, Benjamin E., Joseph A. Allen, Steven G. Rogelberg, Christiane Spitzmüller, Natalie A. DiGiacomo, Jennifer B. Webb, Nathan T. Carter, Olga L. Clark, Lisa A. Teeter, and Alan G. Walker. 2009. Euthanasia-related strain and coping strategies in animal shelter employees. *Journal of the American Veterinary Medical Association* 235(1): 83–88.

Beck, Alan M., and Aaron H. Katcher, 2003. Future directions in human–animal bond research. *American Behavioral Scientist* 47(1): 79–93.

Beerda, Bonne, Matthijs B. H. Schilder, Nicole S. C. R. M. Janssen, and Jan A. Mol. 1996. The use of saliva cortisol, urinary cortisol, and catecholamine measurements for a noninvasive assessment of stress responses in dogs. *Hormones and Behavior* 30(3): 272–279.

Bernard, H. Russell. 2017. *Research methods in anthropology: Qualitative and quantitative approaches.* 6th edition. Baltimore: Rowman & Littlefield.

Best Friends. 2024. No-Kill 2025. https://bestfriends.org/no-kill-2025/animal-shelter-statistics/hawaii.

Blouin, David D. 2013. Are dogs children, companions, or just animals? Understanding variations in people's orientations toward animals. *Anthrozoös* 26(2): 279–294.

Brown, Daniel E. 2016. Stress biomarkers as an objective window on experience. In *Biological Measures of Human Experience across the Lifespan: Making Visible the Invisible.* Lynnette Sievert and Daniel Brown, eds. Pp. 117–141. Cham: Springer. doi:10.1007/978-3-319-44103-0_7.

Bryan, Heather M., Amanda G. Adams, Rosemary M. Invik, Katherine E. Wynne-Edwards, and Judit E. G. Smits. 2013. Hair as a meaningful measure of baseline cortisol levels over time in dogs. *Journal of the American Association for Laboratory Animal Science* 52(2): 189–196.

Cramb, Justin. 2022. The dogs of Remote Oceania: An archaeological and ethnohistorical view of domestic dog introduction and loss in the South Pacific. *Archaeology in Oceania* 57(1): 28–38.

Ebersole, Michele M., and Huihui Kanahele-Mossman. 2020. Broadening understandings of the cultural value of aloha in a teacher educator program. *Journal of Culture and Values in Education* 3(2): 81–99.

Edwards, Christopher, Danielle Giroux, and Scott K. Okamoto. 2010. A review of the literature on Native Hawaiian youth and drug use: Implications for research and practice. *Journal of Ethnicity in Substance Abuse* 9(3): 153–172.

Gray, Peter B., and Sharon M. Young. 2011. Human–pet dynamics in cross-cultural perspective. *Anthrozoös* 24(1): 17–30.

Guruge, Sepali, Lynn Morrison, Vathsala Jayasuriya-Illesinghe, and Theresa Mock. 2016. Intimate partner violence in Hawai'i: Communities in distress. *Arts and Social Sciences Journal* 7(2):e185–e185. doi:10.4172/2151–6200.1000185.

Hawaii Police Department. 2024. 1–11–24 Fentanyl arrests and recovery stats for Dec. 2023. https://www.hawaiipolice.com/1-11-24-fentanyl-arrests-and-recovery-stats-for-dec-2023.

Hawaii State Office of Planning. 2013. https://files.hawaii.gov/dbedt/op/gis/maps/ACS2011_Poverty_Persons.pdf.

Hennessy, Michael B., Victoria L. Voith, Samuel J. Mazzei, Jeff Buttram, Deborah D. Miller, and Fran Linden. 2001. Behavior and cortisol levels of dogs in a public animal shelter, and an exploration of the ability of these measures to predict problem behavior after adoption. *Applied Animal Behaviour Science* 73(3): 217–233.

Heimbürge, Susen, Ellen Kanitz, and Winfried Otten. 2019. The use of hair cortisol for the assessment of stress in animals. *General and Comparative Endocrinology* 270:10–17.

Kessler, Lawrence Helfgott. 2016. *Planter's paradise: Nature, culture, and Hawaii's sugarcane plantations.* Ph.D. dissertation. Temple University, Philadelphia.

Kim, Jaemin, Falina J. Williams, Dayna L. Dreger, Jocelyn Plassais, Brian W. Davis, Heidi G. Parker, and Elaine A. Ostrander. 2018. Genetic selection of athletic success in sport-hunting dogs. *Proceedings of the National Academy of Sciences* 115(30):E7212–E7221.

Krishnamurti, Lauren Sealy. 2021. The potential of "watchful" care: Preventing suicide with aloha in Hawaii. *Medical Anthropology Quarterly* 35(1): 120–135.

Lamon, Tennille K., Margaret R. Slater, Heather K. Moberly, and Christine M. Budke. 2021. Welfare and quality of life assessments for shelter dogs: A scoping review. *Applied Animal Behaviour Science* 244: 105490. doi:10.1016/j.applanim.2021.105490.

Levin, Jack, Arnold Arluke, and Leslie Irvine. 2017. Are people more disturbed by dog or human suffering?: Influence of victim's species and age. *Society & Animals* 25(1): 1–16.

Lohmann, Roger. 2011. Empathic perception and imagination among the Asabano: Lessons for anthropology. In *The anthropology of empathy: Experiencing the lives of others in Pacific societies.* Douglas W. Hollan and C. Jason Throop, eds. Pp. 95–116. New York: Berghan Books. doi:10.1515/9780857451033–006.

Losey, Robert J., Sandra Garvie-Lok, Jennifer A. Leonard, M. Anne Katzenberg, Mietje Germonpré, Tatiana Nomokonova, Mikhail V. Sablin, Olga I. Goriunova, Natalia E. Berdnikova, and Nikolai A. Savel'ev. 2013. Burying dogs in ancient Cis-Baikal, Siberia: Temporal trends and relationships with human diet and subsistence practices. *PLoS One* 8(5):e63740. doi:10.1371/journal.pone.0063740

Luomala, Katharine. 1960. A history of the binomial classification of the Polynesian native dog. In *Culture in history: Essays in honor of Paul Radin.* Stanley Diamond, ed. Pp. 190–240. New York: Columbia University.

McConnell, Allen R., Christina M. Brown, Tonya M. Shoda, Laura E. Stayton, and Colleen E. Martin. 2011. Friends with benefits: On the positive consequences of pet ownership. *Journal of Personality and Social Psychology* 101 (6): 1239–1252.

McNicholas, June, Andrew Gilbey, Ann Rennie, Sam Ahmedzai, Jo-Ann Dono, and Elizabeth Ormerod. 2005. Pet ownership and human health: A brief review of evidence and issues. *BMJ* 331(7527): 1252–1254.

Mishra, Shreya, and Ajoy K. Dey. 2022. Understanding and identifying "themes" in qualitative case study research. *South Asian Journal of Business and Management Cases* 11(3): 187–192.

Morell, Virginia. 2015. From wolf to dog. *Scientific American* 313(1): 60–67.

Morrison, Lynn, Julie Ann Luiz Adrian, Marina Kelley, Johana Hill, Zachariah Tman, and Dana-lynn Ko'omoa-Lange. 2021. Attitude toward companion and guard dogs in Hawaii: Health and welfare implications. *People and Animals: The International Journal of Research and Practice* 4(1): 5. https://docs.lib.purdue.edu/paij/vol4/iss1/5.

Morrison, Lynn. 2023. It's not paradise for the dogs and shelter workers: Dog welfare and occupational stress in animal shelters in Hawaii. *People and Animals: The International Journal of Research and Practice* 6(1): 11. https://docs.lib.purdue.edu/paij/vol6/iss1/11.

Ouschan, Claudia, Alexandra Kuchar, and Erich Möstl. 2013. Measurement of cortisol in dog hair: A noninvasive tool for the diagnosis of hypercortisolism. *Veterinary Dermatology* 24(4): 428–431, e93–e94.

Pukui, Mary Kawena, and Samuel H. Elbert. 1986. *Hawaiian dictionary: Hawaiian-English English-Hawaiian revised and enlarged edition.* Honolulu: University of Hawaii Press.

QSR International Pty Ltd. 2018. NVivo Qualitative data analysis software. In *Statista Global #1 Business Data Platform* (Version 10). https://www.statista.com/statistics/198100/dogs-in-the-united-states-since-2000/.

Rogelberg, Steven G., Natalie DiGiacomo, Charlie L. Reeve, Christiane Spitzmüller, Olga L. Clark, Lisa Teeter, Alan G. Walker, Nathan T. Carter, and Paula G. Starling. 2007. What shelters can do about euthanasia-related stress: An examination of recommendations from those on the front line. *Journal of Applied Animal Welfare Science* 10(4): 331–347.

Shipman, Pat. 2021. *Our Oldest Companions: The Story of the First Dogs.* Cambridge, MA: Harvard University Press.

Schöberl, Iris, Manuela Wedl, Barbara Bauer, Jon Day, Erich Möstl, and Kurt Kotrschal. 2012. Effects of owner–dog relationship and owner personality on cortisol modulation in human–dog dyads. *Anthrozoös* 25(2): 199–214.

Strauss, Anselm, and Juliet Corbin. 1997. *Grounded theory in practice.* Thousand Oaks: Sage.

Taylor, Steven J., Robert Bogdan, and Marjorie L. DeVault. 2015. *Introduction to qualitative research methods: A guidebook and resource.* Hoboken, NJ: John Wiley & Sons.

Titcomb, Margaret. 1969. *Dog and man in the ancient Pacific with special attention to Hawaii.* Honolulu: Bernice P. Bishop Museum.

Uchima, Olivia, Yan Yan Wu, Colette Browne, and Kathryn L. Braun. 2019. Disparities in diabetes prevalence among Native Hawaiians/Other Pacific Islanders and Asians in Hawai'i. *Preventing chronic disease* 16:E22. doi:10.5888/pcd16.180187.

Wright, Kathy D., Ronald Hickman, and Mark L. Laudenslager. 2015. Hair cortisol analysis: A promising biomarker of HPA activation in older adults. *Gerontologist* 55:S140–145.

World Population Review. 2024. Population of Counties in Hawaii. https://worldpopulationreview.com/states/hawaii/counties.

Wynne, Clive. 2019. *Dog is love: Why and how your dog loves you.* Boston: Houghton Mifflin.

5 Is a Dog an Artifact? Archaeological Investigations of Dogs and Their Humans in Oceania

Karen Greig

Introduction

Dogs have been part of human communities for thousands of years and have become entangled in many ways in human cultures around the world. This diversity extends to the place of dogs in Oceanic cultures, where dogs and dog populations are variously owned, managed, cared for, tolerated, abandoned, or otherwise. On some Oceanic islands people's interactions with dogs go back millennia, while on others the relationship is a relatively recent one. Unlike movements across continental landmasses, dogs must have been transported by their people across the waters that separate islands. Oral histories, traditions, contemporary practices and ethnographic accounts attest to a diverse range of interactions between dogs and people, giving an insight into dog–human relationships in the recent past (e.g. Baldwin 1990; Luomala 1960; Titcomb 1969). Dogs are variously companions, pets, hunting aides, watch dogs, bed warmers, camp cleaners, and are often part of the tapestry of village life. These human–canid relationships exist through connections that are forged between dogs and people as living beings. Archaeology provides a unique source of information about dogs and their people further removed in time, however, the nature of this information and how it may be understood differs from more recent accounts.

Broadly speaking, archaeology is the study of human history from its material remains, a "discipline of things" (Olsen et al. 2012). Many of these small, archaeological things are artifacts, defined here as portable objects made by people in the past. There are items made from the bones and teeth of dogs found in archaeological sites across Oceania which readily fit this definition, such as fishhooks, tools, and dog bone and tooth beads (e.g. Davidson 1987; Howes, Jones, and Spriggs 2022; McNiven 2022). Yet there are other material remains of dogs in sites that vary in form and function, including unmodified, disarticulated skeletal material, bones with cut marks from butchery practices, and deliberate dog burials. There is an ambiguity that exists with regards to animals, that Serpell (2009:633) describes as occupying an intermediate place between the

DOI: 10.4324/9781003658696-5

"world of human subjects and the world of insensate objects or things". How then do these bodily remains of dogs fit within archaeological narratives about the past? Do archaeological interpretations of dogs in Oceania capture the diversity of the relationships between people and dogs that are present in contemporary practices and ethnographic accounts? The purpose of this chapter is to examine the predominant archaeological approaches to the study of dogs in Oceania, highlight areas of current research, and explore some emerging ways of thinking about animal–human relationships that may enrich archaeological interpretations of dogs in the past. In doing so, I consider how the remains of dogs' bodies recovered from archaeological sites have been viewed through the lens of human intent and suggest an alternative perspective that archaeological dogs are more than simply artifacts of human behaviour and can be seen to represent not only aspects of human cultures, but of human–canine interspecies cultures as well.

Dogs in Oceania

Although the dog has one of the oldest and most complex relationships with people (Reitz and Wing 2008), evidence of dogs appears relatively late in the human history of Oceania. Recent advancements in archaeology, biological anthropology, and historical linguistics have produced a broad understanding of the deep history of the region, no longer predicated on the 19th century racial divisions of Melanesia, Micronesia, and Polynesia (Kirch 2010). Instead, Oceania is divided between Near and Remote Oceania, reflecting differences in human settlement history and island biogeography. The islands of New Guinea, the Bismarck Archipelago, and Solomon Islands extend across the area defined as Near Oceania, and to the north, east, and southwest lies Remote Oceania, a vast area of ocean interspersed with relatively smaller islands with depauperate biodiversity (Green 1991). The older divisions retain some currency as geographical descriptors, particularly the Polynesian triangle, which extends from Samoa and Tonga (West Polynesia) to its eastern tips, Hawai'i, Rapa Nui (Easter Island), and Aotearoa New Zealand.

The initial arrival of people to the larger landmasses of Near Oceania occurred in the late Pleistocene, with humans reaching continental Australia and New Guinea around 45,000 to 50,000 years ago (O'Connell et al. 2018). These first groups of people relied on endemic fauna and flora, while the introduction of domesticated animals (including dogs) occurred much later. Compared with the antiquity of human occupation in Near Oceania, the islands of Remote Oceania were not colonized by people until around 3,300 years ago (Kirch 2017). This process commenced during the migrations of people identified in the archaeological record by the appearance of a distinctive form of pottery, known as "Lapita", which is present in numerous sites that extend from Papua New Guinea, through the Bismarck

Archipelago, Solomon Islands, Vanuatu, New Caledonia, and Fiji to the western margins of Polynesia (spanning the Near and Remote Oceanic boundary). The introduction of domesticated animals—dogs, pigs, and chickens—to Oceania is associated with these migrations (Kirch 2017). The settlement of Polynesia occurred in two phases, beginning with the Tongan and Samoan islands of West Polynesia around 3,000 years ago during the Lapita expansion (Kirch 2017). Then, after a substantial hiatus, East Polynesia was rapidly settled over several hundred years commencing with the arrival of people to the central tropical islands about 1,000 years ago (Wilmshurst et al. 2011). This second settlement phase included the islands at the margins of East Polynesia: Hawaii, Rapa Nui, and Aotearoa New Zealand.

Documentation about the presence (or absence) of dog remains found in archaeological sites across the Pacific is interspersed throughout the archaeological literature; Cramb (2022) has noted the challenges of working with this scattered information. Recent syntheses highlight the variability of dog introductions and extirpations through time based on reported data from archaeological excavations (Cramb 2022; Greig et al. 2016; Manne et al. 2020; Specht et al. 2014; Sutton et al. 2009; Wickler 2004; see also Koungoulos, this volume, Chapter 7). Archaeological evidence for dogs in Near Oceania appears in Lapita archaeological sites in the Bismarck Archipelago, although in very low frequencies (Specht et al. 2014), with the earliest occurrence in the early Lapita layers of the Kamgot site on Anir Island dating to c. 3,000 to 3,300 years ago. The first evidence of dogs on mainland New Guinea comes from late Lapita sites at Caution Bay on the south coast, which date to between 2,500 to 2,700 years ago (Manne et al. 2020). Notably, dog skeletal remains have not been reported from any Lapita or later archaeological sites in Vanuatu or New Caledonia (Bedford 2006; Sand 2000) creating the appearance of discontinuity between Near and Remote Oceania. A recent parasitological study has however detected helminth parasite eggs from the dog nematode *Toxocara canis* in soil samples from three sites from Vanuatu that pre-date European contact (Horrocks et al. 2024a), as well as at the Kamgot Lapita site (Horrocks et al. 2024b), suggesting the presence of dogs in the landscape, even if not incorporated into the archaeological record.

The later post-Lapita settlement of the islands further to the north and east of Vanuatu and New Caledonia did involve the transport of dogs, but their remains are not uniformly distributed on islands and island groups (Anderson 2009). Dogs are first documented in the archaeological record of central-eastern Micronesia around 2,000 years ago, however, their distribution is uneven across the rest of this area (Wickler 2004). Archaeological evidence for dogs in Polynesia is also variable (Cramb 2022). There is no evidence of dogs in Lapita sites in Tonga and Samoa to the west, where they do not appear in the archaeological record until about 1,000 years ago (Smith 2002). Dogs are also absent from Rapa Nui throughout the pre-colonial period. This contrasts with many of the island groups of East Polynesia, such as Hawaiʻi and New Zealand, where large numbers of dogs are found in early sites and continue to be present throughout the sequence.

The processes of dog domestication and dispersal worldwide remain a much-debated topic, with the application of molecular genetic approaches, including ancient DNA, providing new and much more detailed information about dog ancestry and evolution than was previously available from the macroscopic study of dog bone (Franz et al. 2020; Zhang et al. 2020a). Molecular data indicate that the indigenous dogs of Oceania are descended from domestic Mainland Southeast Asian dogs (Bergström et al. 2020; Greig et al. 2018b; Zhang et al. 2020b). Several mitochondrial lineages have been identified in archaeological and modern dog specimens from Papua New Guinea, with the genetic relationships between Australian dingoes, New Guinea Singing Dogs, modern village dogs, and archaeological specimens yet to be resolved (Fillios and Taçon 2016). In contrast to the genetic diversity found in Near Oceania, only one mitochondrial lineage has been detected in archaeological specimens from Remote Oceania suggesting a quite different dog dispersal history for this area (Greig et al. 2018a). Today, the dogs of Oceania are a melting pot of indigenous dogs and later introductions from around the world. Recent molecular genetic analysis of South Pacific village dogs found that nearly all have European dog ancestry, despite the presence of indigenous dogs throughout the region at the time of European contact (Shannon et al. 2015). The phenotypes of dogs in Oceania today are even more diverse than the genotypes; the ubiquitous "village dog" appears alongside dogs with a mixture of other characteristics derived from recent modern breed arrivals.

Reconstructing Dogs in the Oceanic Past

Physical evidence for dogs in the deep past in Oceania is largely obtained through the recovery of their skeletal remains from archaeological sites; these are mostly disarticulated and highly fragmentary, rather than complete skeletons. Analyses of dog coprolites recovered from archaeological contexts have also been carried out in Aotearoa New Zealand (Wood et al. 2016). As mentioned above, parasitological studies of soil samples from archaeological sites have reported this form of indirect evidence for the presence of dogs in several sites across tropical Remote Oceania and Aotearoa New Zealand (Horrocks et al. 2024a).

Dog skeletal material is analysed using standardized archaeozoological methods, which involve the identification, where possible, of each bone or bone fragment to element and taxon, along with a count of the number present, as a minimum using the quantification measure NISP (Number of Identified Specimens) (Reitz and Wing 2008). Analytical quantification methods have been much debated in archaeozoological literature, suffice to say that different measures have been developed to reconstruct different components of the animals present (Lyman 2019). For example, relative numbers of dogs present in a site can be estimated using MNI (Minimum Number of Individuals), while meat weights can estimate the protein component that dogs may have contributed to the human diet. The results are usually presented in tabular form showing the numbers and sometimes relative proportions of the different animals present, identified to the lowest taxonomic level possible, along with the results of any other forms of analyses, for example, age estimation, sexing, bone modifications, or pathology. Dog bone is routinely reported from archaeological sites in Oceania in this format, although the amount of dog bones recovered from sites varies hugely across the region.

Intentional dog burials (comprising articulated dog skeletons) have been infrequently reported from Oceanic sites (Morey 2010). The Taurama site in PNG contained a dog buried beneath a post hole (Bulmer 2001). Clark's 1994 survey found evidence of 14 dog burials from archaeological contexts in Polynesia (Clark 1996). Campbell and Hudson (2011) describe an unusual occurrence found during an excavation carried out in South Auckland, Aotearoa New Zealand, where several dogs were found to have been given similar burial treatments to people, in a 17th century Maori burial ground. Using an integrative multi-method approach Kramer and colleagues (2022) investigated aspects of these dogs along with dog bone found disarticulated in nearby middens and found differences in the origins, diet, health, and treatment of the dogs. Dog bones have been recovered elsewhere from archaeological contexts also associated with ritual deposition (e.g. McNiven et al. 2015; Millerstrom 2004).

Archaeological Interpretations

Although not necessarily made explicit, the presentation of dog bones in tables of animal fauna identified from archaeological sites often reflects a focus on the diet and resource use of the site's former inhabitants; animals are viewed in terms of the "protein and calories" they provide (Russell 2012:7). Grahame Clark was an early proponent of research that addressed questions of subsistence and economy, developed during his excavation and analysis of the animal remains from Star Carr (Clark 1954). This research concern was exported to the Pacific by Clark and his students who took up influential teaching positions in Australia and New Zealand in the 1950s (Allen and Nagaoka 2004; Fagan 2018). Clark's (1954) approach examined the local environment in combination with the animal remains to develop an understanding of resource use from the wider area. Fragmentary animal bones are reconstituted as archaeological animals slaughtered for food and consumed as part of everyday sustenance, as part of ritualized practices, or as a source of raw materials.

Research into the use of dogs as a food resource in Polynesia has explored their role in the production of surpluses and possible competition between people and animals. Bay-Petersen (1979, 1983) examined the socio-economic role of dogs and pigs in the pre-contact period and concluded that they were in direct competition with people for agricultural foods, and that the number of animals kept (if any) was the result of a complex balance between human nutritional requirements and the production of surplus from which to feed animals. Pillay and colleagues (2022) recent publication also approaches human-dog relationships in Aotearoa New Zealand from the perspective of interspecific competition, rather than predation or mutualism. Their analysis of dog temporal and geographic distribution patterns and dental markers suggests a period of dog–human competition between 1450 and 1650 CE, which was a period of major environmental, economic, and political change in the country.

Environmental and ecological studies, also drawing on data from animal bones (including dogs) in archaeological sites, have investigated human adaptation to, and impacts on, island environments, particularly on the islands of Remote Oceania. Archaeozoological analyses have been used to document changing proportions of different species through time, changes to island ecosystems, and to highlight animal introductions and extinctions (Kirch 2017). Nagaoka (2002, 2005) analyzed changes in species distribution and frequency of dogs, birds, seals, and fish in Aotearoa New Zealand using behavioral ecology models, including optimal foraging theory, while Anderson (2009) used the concept of biological invasion to explore the distribution of domestic animals, including dogs, in Remote Oceania. Cramb (2022) investigated dog survivorship on islands by analyzing island size, type, and isolation. These kinds of ecological approaches, as with subsistence economics, tend to conceptualize dogs as part of the set of animal resources able to be exploited by people from the wider environment.

Archaeological ideas about subsistence strategies, biogeography, and the manipulation of island ecosystems were brought together in the concept of the "transported landscape", applied to the Pacific region in the 1980s by Kirch (2017). The transported landscape describes a blueprint or cognitive map for the re-creation of a productive environment on newly settled, previously uninhabited islands. It involves the transport of domesticated animals, including dogs, and plant cultigens, along with the knowledge of necessary agricultural practices, which triggered a cascade of changes to island ecosystems increasing in impact across the oceanic islands of Remote Oceania. Kirch argues that the transported landscape of the Lapita people and later their Polynesian descendants was a pivotal strategy in the successful colonization of Remote Oceania. Dogs, along with pigs and chickens, were components of this human strategy for settlement of new islands. While this perspective is not an exclusive way of conceiving of dogs in the past, and there is space for other ways of looking at human–dog interactions, it does foreground the place of dogs as a resource for human communities—one that is subject to their manipulation and control. The inclusion of dogs in these voyages has also contributed to the application of new biomolecular techniques using dog skeletal remains to examine aspects of migration and colonization processes.

Dogs as a Proxy for Humans

In the last few decades, new biomolecular methods using isotopes and ancient DNA have resulted in the use of dog bones in studies of migration, human and animal mobility, trade and exchange, and human diets. The commensal model pioneered by Matisoo-Smith and colleagues addresses questions of human migration, initially using the bones of the Pacific rat (*Rattus exulans*) as a proxy for human remains (Matisoo-Smith 1994, 2009). This approach, now expanded to include modern and ancient DNA of rats, dogs, pigs, and chickens, traces the genetic relationships between animals from various islands and island groups to infer human mobility. The term “commensal” is applied broadly to cover those animals living in close association with humans that were dependent on humans for their dispersal (Matisoo-Smith 2009; Storey et al. 2013). Commensal studies have revealed that the dispersal patterns of dogs, pigs, chickens, and Pacific rats, based on genetic data, vary between species (Collins et al. 2023), and may also differ from the genetic patterning of human founding populations (e.g. Greig et al. 2018a). Ancient DNA analysis of dog bone from archaeological sites (outlined above) reveals a complex history of dog introduction and dispersal in the region, not discernible from standard archaeozoological methods (Greig et al. 2015, 2018a).

As with molecular genetic studies, the analysis of stable isotopes in animal bone has been used to address questions relating to human diets and subsistence (Kinaston and Buckley 2013; Kinaston et al. 2013) and mobility (Shaw et al. 2009). In North American studies using dog bone, this has been termed the Canine Surrogacy Approach (Guiry 2013), making explicit the role of dogs as a surrogate or analogue for their people. There has been limited application of stable isotope methods specifically using dog bone in Oceanic contexts, although researchers working in Polynesia have included some isotopic studies of dog diets in their investigations of human subsistence strategies (e.g. Allen and Craig, 2009; Kinasaton et al. 2013; Richards et al. 2009) and environmental stressors (Littleton et al. 2020).

Archaeological Dogs and Past Human Practices

As part of the first group of introduced Pacific domesticates, dogs are framed as a core component of deliberately translocated animal resources used by people. Dog bones in archaeological sites become artifacts of food preparation and consumption, and manufacturing processes. These fragments of dog bone are small, portable objects produced by human practices in the past, that are the subject of archaeological inquiry in the present. Russell (2012) and Sykes (2014) have critiqued what they argue is an enduring and dominant archaeozoological focus, which privileges the subsistence role of dead animals to the exclusion of others and consequently seriously constrains archaeological interpretations. As Russell (2012:9) observes for animals generally, areas of study that have received less archaeological attention "revolve around the value and meanings of living animals: animal symbolism, animals as wealth, animals as pets". This expanded approach still tends to view dogs from an utilitarian perspective, focusing on the functions that they have in human society. The interactions between dogs and their humans, whether economic, symbolic, or even social, are often framed as the "roles that dogs play" in human societies. Extending this performative metaphor further, is there also an implication that dogs are somehow canine "actors" or "stand ins" for humans? There is often very little consideration of what dog lifeways and behaviour might have been like, and how these may contribute towards the interactions between people and dogs, or with other parts of the environment.

The arrival of pigs, dogs, and chickens has been identified as the first instance of animal husbandry associated with Oceanic forms of agriculture (Spriggs 1996), yet little is known about this form of animal management from an archaeological perspective. Many early ideas about domestication and animal management arose from the changes in peoples' relationships with animals in the Neolithic in continental Eurasia, although multiple domestication centers have now been recognized worldwide (Larson et al. 2014). Contra to that regional perspective, not all forms of animal management relate to increasingly intensive forms of animal husbandry, where returns are maximized through deliberate selection of animals of a particular age and, or sex for culling for consumption. As Keswani (1994) observes, while models that are based on maximising yields from domesticated animals can be useful in the interpretation of state and market centered economies, for other societies, socio-ideological and ritual practices are likely to pattern archaeological assemblages in different ways. A notable feature of Oceanic societies is the incorporation of domesticated animals in social exchange networks, that act to establish and maintain social relationships. Pigs and dogs, alive, as components of feasts, and as a source of valuable items of material culture, have been and still are in some places parts of complex systems of social exchange particularly in New Guinea. The relatively small number of dogs in Near Oceanic archaeological sites especially may not be reflective of their sociocultural importance.

As well as varying between different islands and island groups, the place of dogs in Oceanic societies throughout time cannot be assumed to be static and unchanging. Dogs have been part of human communities in a diverse variety of ways which are not necessarily mutually exclusive, for example, as village dogs, companion animals, and as hunting dogs (Koungoulos and Brumm 2024). Peoples' relationships with dogs have the capacity for change, this is perhaps best illustrated by archaeological evidence for the extirpation of dogs from some islands (Cramb 2022). Conversely, adaptation of animal management practices to long distance voyaging may have enabled dogs (and pigs and chickens) to be first moved into Remote Oceania (Greig et al. 2015).

In the archaeological record, apart from intentional dog burials which imply a bond of some kind between a dog and its humans (Morey 2010), it is challenging to mentally reassemble and reanimate living dogs from collections of fragmentary archaeological dog bones, and to examine aspects of their shared history with people, where they are not simply an artifact of aspects of human behavior, but were active participants in the formation and maintenance of their relationships with people and the wider environment. Emerging archaeological approaches discussed below offer new ways to think about the relationships between dogs and their people in the past; in particular, relationality, social archaeozoology, and multispecies perspectives.

Discussion

Recent trends in the humanities and social sciences concerned with the relationships between humans and non-human animals have emerged as what has been termed the "animal turn" (Ritvo 2007; Boyd 2017). This has foregrounded a recognition that animals as sentient beings possess a subjective form of agency. The philosopher Helen Steward (2009:226) argues for the following concept of agency for animals: an agent can move the whole, or at least part, of its body; it possesses some form of subjectivity; it has at least some types of rudimentary intentional state (such as trying, or wanting), and it is a "settler of matters concerning certain of the movement of its own body … attributed always first and foremost to the agent, and only secondarily to environmental impacts or triggers". Influenced by the concept of relationality (e.g. Ingold 1994, 2000; Watts 2013), archaeologists have explored the ways that animals may have played subjective, agential roles in past societies (e.g. Argent 2010; Armstrong Oma 2010; Hill 2013; Knight 2005; Orton 2010). Animals are conceptualized as living, sentient beings capable of engaging in social interactions with people, rather than being reconstructed as objects, such as food, prey, or sources of raw materials. There are elements of mutual becoming in these relations, as animals' and peoples' behaviour can be formed and reformed through their interactions.

For Hill (2013) relational modes of interaction between people and animals in hunting and gathering societies provide evidence that animals in the past may have been understood as persons. In these societies the boundaries between nature and culture and humans and animals are often permeable, fluid or sometimes even non-existent. Orton (2010) argues that domestic animals can be seen as "sentient property", despite being objects within property relations between people, they are also subjects with their own social world that overlaps with that of people. Like Hill, Orton also points out the limitations of the archaeological and more broadly anthropological viewpoint that adheres to the Cartesian separation between human and non-human. From this viewpoint, animals are seen as one-dimensional objects, rather than "persons" who are capable of entering into intersubjective relationships with individual people. Orton argues for a fundamentally social approach to domestic animals that avoids reducing them to objects; either as material resources or of symbolic or religious importance to past peoples.

Armstrong Oma (2010) frames this engagement between people and domestic animals as a "social contract", with notions of trust and reciprocity at its core. This framing enables animals to have multiple roles and obligations in a community and be simultaneously involved in different areas and activities. Armstrong Oma sees each contract as arising in a particular time and place, rather than being some kind of universally applied concept. In contrast, Lindstrøm (2012) suggests animal burial is a characteristic of a cross-cultural and trans-temporal phenomenon where people experience animals as having personhood and identities. The context and scale of interactions between people and animals will influence the nature of their relationships, with small-scale animal management practices offering more opportunities for the development of close bonds. Even with the potential for close relationships between animals and people, there is still the capacity for domination, just as there is in human relationships (Boyd 2017). The social contract as described by Armstrong Oma (2010) does not imply equal partnerships either, and describes an asymmetrical relationship that could have also varied in the past. Human attitudes towards dogs can be marked by ambivalence, and sometimes cruelty (Lohmann, this volume, Chapter 1).

Around the same time as Hill's (2013) work, Overton and Hamilakis (2013) take issue with the simplistic portrayal of animals as subsistence resources to be used, or symbolic entities for the benefit of humans. They advocate for a social zooarchaeology that "avoids anthropocentrism; begins from a zoontology of species co-shaping; reinstates the agency of animals, both in life and in death; and appreciates the embodied and sensory nature of interspecies interaction". Despite their forceful critique, they do not wish to replace economic, subsistence, or environmental aspects of archaeozoological studies with a new social approach, rather they want archaeozoology to address the complex web of associations that link people and animals across their lifetimes and sometime beyond (2013:135).

There is a challenge therefore for archaeologists working in the Pacific to move beyond utilitarian reconstructions of human–dog relationships in the past, predicated on parts of dog bodies as resources, to explore the possibilities of social and agential nature of human–dog interactions. This endeavor would not detract from the contribution that archaeozoological studies make to important questions about past lifeways in the region, rather (after Overton and Hamilakis 2013; Russell 2012) archaeozoology may also provide a more nuanced understanding of the complex entanglements between animals and humans in the past. In doing so, archaeological studies could produce information that may complement local communities' knowledge about dogs and their cultural meanings, and contribute to the story of human–animal relationships and their place in human societies, from the past to the present day.

Human–animal meeting points are a place of encounter contributing to the formation of relationships, "created by mutual action, becoming and performing" and are a potential area of study of human–animal relationships (Armstrong Oma 2010). In Oceania, the relationships between people and domesticated animals, including dogs, are implicated in significant aspects of archaeological understandings of the past, for example, the transported landscape, complex social exchange systems, and the production of wealth and social differentiation. All of which require interactions between people and living animals. Yet the current interpretations of dog bones simply as artifacts of human subsistence and resource use, or their use as a proxy for human diet or mobility does little to inform on these historical interactions. For example, the ongoing use of dog teeth to create beads dates back several thousand years in Oceania. Holes are drilled in dog canines to create beads or exchange valuables, and these are strung on cords or incorporated into items such as head dresses, necklaces or pectorals (McNiven 2022). Some of the first evidence of dogs in the region come from drilled teeth from Lapita sites in Papua New Guinea (Manne et al. 2020), and they have been recovered from other later archaeological sites (Allen, 1977; Basiaco et al. 2020; Bulmer 1978). Ethnographic accounts document how items made from dogs' teeth were used in social exchange systems, to obtain pigs and as bridewealth (Harding 1994). The importance of dog teeth in these systems was such that in the 19th century ceramic replicas were manufactured in Germany and used as a source of payment in colonial plantations (Wronska-Friend 2015). Some dogs' teeth necklaces in museum collections comprise a mix of real and these replica teeth. As a dog only possesses four canine teeth, the number of dogs involved in the production of some of these items would be considerable. What are the implications of this practice for dog–human interactions in the past, to what extent did people influence or manage dog populations to produce these valuables? Harding (1994) also documents the trading of live dogs, which could also have contributed to the process of dog dispersal across the region in the past.

The incorporation of dog bones into archaeological contexts occurs at an ending of one aspect of the relationship between people and dogs. Yet those dogs have previously shared domestic space with people, would have interacted with people over their lifetime, and possess their own life histories. People may have manipulated aspects of dogs' lifeways, and both may have responded to the presence and activities of the other in their shared environment. For example, as living sentient beings, puppies must socialize themselves with humans and other animals in their environment (Armstrong Oma and Birke 2013). Enculturation of dogs and their humans can be a mutual process (Lohmann, this volume Chapter 8). Just as dogs become socialized, people may have adapted their behavior to contend with aspects of life with dogs. Shared living can have negative as well as positive or neutral consequences, and this can include injuries from dog bites, competition for food and predation on other livestock (e.g. Chowning 1991; Feinberg and Pyrek, this volume, Chapter 6; Prokosch, this volume, Chapter 3). Being near dogs may have also exposed people to disease and parasites, where dogs either suffer from the disease or are carriers. There may also be negative consequences from dogs' interactions with non-domestic animals (Greig and Rawlence 2021). Dog behavior may have also influenced aspects of the archaeological record. Perhaps the most well-known example of this is the consumption of food waste by dogs, including the gnawing of bones, digging, and other ground disturbance (Spennemann 1990). And some archaeological evidence associated with dogs in archaeological sites may be just that, evidence about dogs and dog behavior independent of their humans.

If we understand human–animal relationships as part of a wider network of connections between people, animals, objects, and places, then the emerging interdisciplinary field of multispecies archaeology may provide a useful framework to examine human–animal interactions in the Oceanic past. In the edited volume of the same name, editor Pilaar Birch (2018) sets out an approach where humans are no longer the subject of the question, rather they are instead treated as one actor (agent) among others. Birch's approach is strongly influenced by ecology, to the extent that she suggests it could be viewed as "archaeo-ecology"; a shared history of life, where the past is understood through a series of networks and interactions. Boyd however questions whether attempts to address anthropocentrism in anthropological research more broadly may go too far, that the "move to decenter the human, while laudable, can tend to erase our own situatedness both as human scholars and as the dominant force of change on the planet" (Boyd 2017:303).

Littleton and colleagues review the potential of a multispecies approach in Remote Oceania (2020). They demonstrate the application of isotopes, dental pathology, and enamel hypoplasia analyses to identify major environmental stressors and how these affected individual species, including dogs, before examining possible changes in animal husbandry methods by people as a response. In Aotearoa New Zealand, differences between the genetic diversity of founding populations of people and their dogs indicate that the commensal model has the potential to be informative about the process of dog introduction and establishment, in addition to the human colonisation process (Greig et al. 2018a). Furthermore, the pattern of dog introduction and dispersal in Oceania, involving multiple mitochondrial genetic lineages (Greig et al. 2018b), contrasts with that of the single Pacific clade found in pigs (Larson et al. 2007; Horsburgh et al. 2022). What do these differences mean for understanding species-specific human and animal interactions? Other studies foregrounding questions about human lifeways that incorporate dog and other animal proxies, such as isotopic dietary studies, also have the potential to inform on aspects of dogs' lives in their own right.

In contrast to Birch's (2018) ecological conception of multispecies archaeology, Živaljević (2021) offers a more socially oriented perspective, advocating for a multispecies archaeology that "paves the way to immerse or knowledges of various organisms which coexisted with humans in questions of materiality, agency, personhood, embodiment and entanglement". Conceptualizing animals as persons recognizes the formation and maintenance of kinship and other social ties between people and dogs (and other animals) found throughout many communities in Oceania (e.g. Chowning 1991; Dobrin, this volume, Chapter 2). Furthermore, these relationships can continue after the animals have died, through cosmological relationships and embedded in objects made from their bones (Tsang et al. 2022). This social approach has parallels with the interspecies cultures explored throughout this volume. Lohmann (this volume, Chapter 1) sees dog phenotypes as an outcome of "selective breeding for ideals … which are manifestations of human culture that require canine bodies to be expressed". Culture consists of creations that exist in three media: thoughts, behaviors, and artifacts, and it is through bodily expression or material culture that these creations are transmitted to others (Lohmann 2010). Dogs can be seen as a form of cultural creation, a product of mind (and action). More than insensate artifacts, they become an animated and embodied expression of culture.

To conclude, dogs and people in Oceania have shared social and physical spaces, in some places for several thousand years. As living, sentient beings, dogs have interacted with each other, with members of their human community, and with the world around them. Dogs are capable of influencing their world, their relationships with people and other animals, and even the formation and transformation of the archaeological record itself. These interactions are historically situated, some will play out within the lifetime of one dog while others may occur over long periods of time in a group of dogs. While many of these interactions leave little or no trace in the archaeological record, others will. Where these traces are embodied in the remains of dog bones, these bones become more than artifacts or objects of human culture, rather they are expressions of human–canine interspecific culture shared by people and dogs. Recognizing the potential for multispecies approaches that encompass relationality, diverse worldviews, interspecies cultures, and the agency of dogs offers new opportunities for archaeozoologists to explore the ways in which dogs were entangled in the lifeways of humans and other animals, and how these relationships may be manifest in the material remains of Pacific pasts.

Acknowledgements

Many thanks to Roger Lohmann and an anonymous reader for their insightful comments that greatly improved this chapter. By necessity, it is a selective survey, rather than a detailed account of dogs in the archaeological literature, my apologies for any omissions. Support from the Royal Society of New Zealand Te Apārangi Marsden Fund MFP-UOO2020 is gratefully acknowledged.

References

Allen, Jim. 1977. Fishing for wallabies: Trade as a mechanism for social interaction, integration and elaboration on the central Papuan coast. In *The evolution of social systems.* J. Friedman and M. J. Rowlands, eds. Pp. 419–455. London: Duckworth.

Allen, Melinda S., and Jacqueline A. Craig. 2009. Dynamics of Polynesian subsistence: Insights from archaeofauna and stable isotope studies, Aitutaki, Southern Cook Islands. *Pacific Science* 63(4): 477–506.

Allen, Melinda S., and Lisa Nagaoka. 2004. "In the footsteps of von Haast … the discoveries something grand": The emergence of zooarchaeology in New Zealand. In *Change through time: 50 years of New Zealand archaeology.* Louise Furey and Simon Holdaway, eds. Pp. 193–214. Auckland: Archaeological Association Monograph.

Anderson, Atholl. 2009. The rat and the octopus: Initial human colonization and the prehistoric introduction of domestic animals to Remote Oceania. *Biological Invasions* 11(7): 1503–1519.

Argent, Gala. 2010. Do the clothes make the horse? Relationality, roles and statuses in Iron Age Inner Asia. *World Archaeology* 42(2): 157–174.

Armstrong Oma, Kristin. 2010. Between trust and domination: Social contracts between humans and animals. *World Archaeology* 42(2): 175–187.

Armstrong Oma, Kristin, and Lynda Birke. 2013. Guest Editors' Introduction Archaeology and Human–Animal Studies. *Society & Animals* 21(2): 113–119.

Baldwin, James A. 1990. Muruk, dok, pik, kakaruk: Prehistoric implications of geographical distributions in the southwest Pacific. In *Pacific Production systems: Approaches to economic prehistory. Papers from a Symposium at the XV Pacific Science Congress, Dunedin, New Zealand 1983.* D. E. Yen and J. M. Mummery, eds. Canberra, ACT: Dept. of Prehistory, Research School of Pacific Studies, The Australian National University.

Basiaco, Adriana, Chris Urwin, and Tiina Manne. 2020. Worked bone and teeth from Orokolo Bay in the Papuan Gulf (Papua New Guinea). *Australian Archaeology* 86(3): 226–237.

Bay-Petersen, Jan Allo. 1979. The role of the dog in the economy of the New Zealand Maori. In: *Birds of a Feather: Osteological and archaeological papers from the South Pacific in honour of RJ Scarlett. New Zealand Archaeological*

Association Monograph, 11. Pp 165–181. Dunedin, NZ: New Zealand Archaeological Association.

Bay-Petersen, Jan. 1983. Competition for resources: The role of pig and dog in the Polynesian agricultural economy. *Journal de la Société des Océanistes* 39(77): 121–129.

Bedford, S. 2006. Pieces of the Vanuatu puzzle: Archaeology of the north, south and centre. *Terra Australis 23*. Canberra: Pandanus Books.

Bergström, Anders, Laurent Frantz, Ryan Schmidt, Erik Ersmark, Ophelie Lebrasseur, Linus Girdland-Flink, Audrey T. Lin, Jan Storå, Karl-Göran Sjögren, and David Anthony. 2020. Origins and genetic legacy of prehistoric dogs. *Science* 370(6516): 557–564.

Birch, Suzanne E Pilaar, ed., 2018. *Multispecies archaeology*. New York: Routledge.

Boyd, Brian. 2017. Archaeology and human–animal relations: Thinking through anthropocentrism. *Annual Review of Anthropology* 46(1): 299–316.

Bulmer, Susan. 1978. *Prehistoric culture change in the Port Moresby region*. Ph.D thesis, University of Papua New Guinea.

Bulmer, Susan. 2001. Lapita dogs and singing dogs and the history of the dog in New Guinea. In *The archaeology of Lapita dispersal in Oceania: Papers from the Fourth Lapita Conference, June 2000, Canberra, Australia*. G. R. Clark, A. J. Anderson, and T. Vunidilo, eds. Pp 183–201. Canberra: Pandanus Books.

Campbell, Matthew, and Beatrice Hudson. 2011. *The NRD site community report*. Auckland, New Zealand: Unpublished report, CFG Heritage.

Chowning, Ann. 1991. Pigs, dogs and children in Molima. In *Man and a Half: Essays in Pacific Anthropology and Ethnobiology in Honour of Ralph Bulmer*. Andrew Pawley, ed. Pp. 182–187. Auckland: The Polynesian Society.

Clark, Geoffrey. 1996. Animal burials from Polynesia. *Archaeology in New Zealand* 39: 30–38.

Clark, Grahame. 1954. *Excavations at Star Carr: An early mesolithic site at Seamer near Scarborough, Yorkshire*. Cambridge: Cambridge University Press.

Collins, Catherine, Anna L. Gosling, and Elizabeth Matisoo-Smith. 2023. Commensal Models in Island Archaeology: Animal and Plant Translocations. In *The Oxford Handbook of Island and Coastal Archaeology*. Scott M. Fitzpatrick and Jon Erlandson, eds. Oxford: Oxford University Press.

Cramb, Justin. 2022. The Dogs of Remote Oceania: An archaeological and ethnohistorical view of domestic dog introduction and loss in the South Pacific. *Archaeology in Oceania* 57(1): 28–38.

Davidson, Janet. 1987. *The prehistory of New Zealand*. Auckland: Longman Paul.

Fagan, Brian. 2018. *Grahame Clark: An intellectual life of an archaeologist*. Boulder, CO: Westview Press.

Fillios, Melanine A., and Paul S. Taçon. 2016. Who let the dogs in? A review of the recent genetic evidence for the introduction of the dingo to Australia and implications for the movement of people. *Journal of Archaeological Science: Reports* 7: 782–792.

Frantz, Laurent A. F., Daniel G. Bradley, Greger Larson, and Ludovic Orlando. 2020. Animal domestication in the era of ancient genomics. *Nature Reviews Genetics* 21(8): 449–460.

Green, R. C. 1991. Near and Remote Oceania: Disestablishing 'Melanesia' in culture history. In *Man and a half: Essays in Pacific anthropology and ethnobiology in honour of Ralph Bulmer*. A. Pawley, ed. Pp. 491–501. Auckland: The Polynesian Society.

Greig, Karen, James Boocock, Stefan Prost, K. Ann Horsburgh, Chris Jacomb, Richard Walter, and Elizabeth Matisoo-Smith. 2015. Complete Mitochondrial Genomes of New Zealand's First Dogs. *PLoS One* 10(10):e0138536.

Greig, Karen., Richard Walter, and Elizabeth Matisoo-Smith. 2016. Dogs and people in Southeast Asia and the Pacific. In *The Routledge Handbook of Bioarchaeology in Southeast Asia and the Pacific Islands.* Marc Oxenham, and Haillie Buckley, eds, pp. 462–482. London and New York: Routledge.

Greig, Karen, James Boocock, Melinda S. Allen, Richard Walter, and Elizabeth Matisoo-Smith. 2018a. Ancient DNA evidence for the introduction and dispersal of dogs in New Zealand. *Journal of Pacific Archaeology* 9(1): 1–10.

Greig, Karen, Anna Gosling, Catherine J. Collins, James Boocock, K. McDonald, D. J. Addison, Melinda S. Allen, Bruno David, M. Gibbs, C. F. W. Higham, F. Liu, Ian J. McNiven, Sue O'Connor, C. H. Tsang, Richard Walter, and Elizabeth Matisoo-Smith. 2018b. Complex history of dog (*Canis familiaris*) origins and translocations in the Pacific revealed by ancient mitogenomes. *Scientific Reports* 8(1): 9130.

Greig, Karen, and Nicolas J. Rawlence. 2021. The contribution of kurī (Polynesian dog) to the ecological impacts of the human settlement of Aotearoa New Zealand. *Frontiers in Ecology and Evolution* 9: 757988.

Guiry, Eric J. 2013. A canine surrogacy approach to human paleodietary bone chemistry: Past development and future directions. *Archaeological and Anthropological Sciences* 5: 275–286.

Harding, Thomas G. 1994. Precolonial New Guinea trade. *Ethnology* 33(2): 101–125.

Hill, Erica. 2013. Archaeology and Animal Persons: Toward a Prehistory of Human-Animal Relations. *Environment and Society: Advances in Research* 4(1): 117–136.

Horrocks, Mark, Stuart Bedford, James Flexner, Frédérique Valentin, Ethan Cochrane, Melina S. Allen, David Burley, Patrick V. Kirch, and Bronwen Presswell. 2024a. New dog parasite (*Toxocara canis*) discoveries at Melanesian and Polynesian sites in tropical Remote Oceania. *Pacific Science* 78(1): 19–30.

Horrocks, Mark, Glenn Summerhayes, and Bronwen Presswell. 2024b. Palaeoparasitology confirms Early Lapita evidence of pig and dog at Kamgot, Bismarck Archipelago. *Journal of Helminthology* 98:e8.

Horsburgh, K. Ann, Anna L. Gosling, Ethan E. Cochrane, Patrick V. Kirch, Jillian A. Swift, and Mark D. McCoy. 2022. Origins of Polynesian pigs revealed by mitochondrial whole genome ancient DNA. *Animals* 12(18): 2469.

Howes, Hilary, Tristen Jones, and Matthew Spriggs, eds. 2022. *Uncovering Pacific Pasts: Histories of Archaeology in Oceania*. Canberra: ANU Press.

Ingold, T. 1994. From Trust to Domination: An Alternative History of Human-Animal Relations. In *Animals and Human Society.* Aubrey Manning and James Serpell, eds. Pp. 1–22. London: Routledge.

Ingold, T. 2000. *The Perception of the Environment*. London: Routledge.

Keswani, Priscilla. S. 1994. The social context of animal husbandry in early agricultural societies: Ethnographic insights and an archaeological example from Cyprus. *Journal of Anthropological Archaeology* 13: 255–277.

Kinaston, Rebecca. L., and Hallie. R. Buckley, 2013. The stable isotope analysis of prehistoric human diet in the Pacific Islands with an emphasis on Lapita. In *Pacific archaeology: Documenting the past 50,000 years: Papers from the 2011 Lapita Pacific Archaeology Conference.* Glenn Summerhayes and Hallie

Buckley, eds. Pp. 91–106. University of Otago Studies in Archaeology. Dunedin, New Zealand: University of Otago Press.

Kinaston, Rebecca L., Richard K. Walter, Chris Jacomb, Emma Brooks, Nancy Tayles, Sian E. Halcrow, Claudine Stirling, Malcolm Reid, Andrew R. Gray, and Jean Spinks. 2013. The first New Zealanders: Patterns of diet and mobility revealed through isotope analysis. *PloS one* 8(5):e64580.

Kirch, Patrick. 2010. Peopling the Pacific: A holistic anthropological perspective. *Annual Review of Anthropology* 39: 131–148.

Kirch, Patrick V. 2017. *On the road of the winds: An archaeological history of the Pacific Islands before European contact.* Oakland, CA: University of California Press.

Koungoulos, Loukas G. and Adam Brumm. 2024. Hunting with dogs: A synthesis of ethnohistorical data and discussion of their implications for prehistoric subsistence in New Guinea. *Archaeology in Oceania* 59(1): 1–28.

Kramer, Robyn T., Karen Greig, Matthew Campbell, Patricia Pillay, Melinda Allen, Charlotte King, Hallie Buckley, Clément Bataille, Beatrice Hudson, and Stuart Hawkins. 2022. An Exploratory and Integrative Study of Māori Kurī (*Canis familiaris*) at the NRD Archaeological Site in Aotearoa New Zealand. *Journal of Pacific Archaeology* 13(1): 8–27.

Larson, Greger, Thomas Cucchi, Masakatsu Fujita, Elizabeth Matisoo-Smith, Judith Robins, Atholl Anderson, Barry Rolett, Matthew Spriggs, Gaynor Dolman, Tae-Hun Kim, Nguyen Thi Dieu Thuy, Ettore Randi, Moira Doherty, Rokus Awe Due, Robert Bollt, Tony Djubiantono, Bion Griffin, Michiko Intoh, Emile Keane, Patrick Kirch, Kuang-Ti Li, Michael Morwood, Lolita M. Pedriña, Philip J. Piper, Ryan J. Rabett, Peter Shooter, Gert Van den Bergh, Eric West, Stephen Wickler, Jing Yuan, Alan Cooper, and Keith Dobney. 2007. Phylogeny and ancient DNA of Sus provides insights into neolithic expansion in Island Southeast Asia and Oceania. *Proceedings of the National Academy of Sciences* 104(12): 4834–4839.

Larson, Greger, Dolores R. Piperno, Robin G. Allaby, Michael D. Purugganan, Leif Andersson, Manuel Arroyo-Kalin, Loukas Barton, Cynthia Climer Vigueira, Tim Denham, and Keith Dobney. 2014. Current perspectives and the future of domestication studies. *Proceedings of the National Academy of Sciences* 111(17): 6139–6146.

Lindstrøm, Torill Christine. 2012. "I am the walrus": Animal identities and merging with animals – exceptional experiences? *Norwegian Archaeological Review*, doi:10.1080/00293652.2012.703687.

Littleton, Judith, Gina McFarlane, and Melinda S. Allen. 2020. Human–animal entanglements and environmental change: Multi-species approaches in Remote Oceania. In *The Routledge Handbook of the Bioarchaeology of Climate and Environmental Change.* Gwen Schug, ed. Pp. 493–510. London and New York: Routledge.

Lohmann, Roger Ivar. 2010. Introduction: The Anthropology of Creations. *Anthropological Forum* 41(2): 97–104.

Luomala, Katherine. 1960. The native dog in the Polynesian system of values. In: *Culture in history: Essays in honor of Paul Rodin.* Stanley Diamond, ed. Pp. 190–240. New York: Columbia University Press.

Lyman, R. Lee. 2018. A Critical Review of Four Efforts to Resurrect MNI in Zooarchaeology. *Journal of Archaeological Method and Theory* 26(1): 52–87.

Manne, Tiina, Bruno David, Fiona Petchey, Matthew Leavesley, Georgia Roberts, Katherine Szabó, Chris Urwin, Ian J. McNiven, and Thomas Richards. 2020. How long have dogs been in Melanesia? New evidence from Caution Bay, south coast of Papua New Guinea. *Journal of Archaeological Science: Reports* 30:102255.

Matisoo-Smith, Elizabeth. 1994. The human colonisation of Polynesia—A novel approach: Genetic analyses of the Polynesian rat (Rattus exulans) . *Journal of the Polynesian Society* 103: 75–87.

Matisoo-Smith, Elizabeth. 2007. Animal translocations, genetic variation and the human settlement of the Pacific. In *Genes, language and culture history in the Southwest Pacific*. J. S. Friedlaender, ed. Pp. 157–170. Oxford: Oxford University Press.

Matisoo-Smith, Elizabeth. 2009. The Commensal Model for human settlement of the Pacific 10 Years on—What can we say and where to now? *The Journal of Island and Coastal Archaeology* 4: 151–163.

McNiven, Ian J., Duncan Wright, Shannon Sutton, Marshall Weisler, Scott Hocknull, and John Stanisic. 2015. Midden formation and marine specialisation at Goemu village, Mabuyag, Torres Strait, before and after European contact. *Memoirs of the Queensland Museum, Culture* 8(2): 377–475.

McNiven, Ian J. 2022. Beyond bridge and barrier: Reconceptualising Torres Strait as a coconstructed border zone in ethnographic object distributions between Queensland and New Guinea. *Queensland Archaeological Research* 25: 25–46.

Millerstrom, Sidsel N. 2004. Facts and fantasies: The archaeology of the Marquesan dog. In *Behaviour behind bones: The zooarchaeology of ritual, religion, status and identity.* Sharyn Jones O'Day, Anton Ervynck, and Wim Van Neer, eds. Pp. 144–152. Oxford: Oxbow Books.

Morey, Darcy F. 2010. *Dogs.* Cambridge: Cambridge University Press.

Nagaoka, Lisa. 2002. Explaining subsistence change in southern New Zealand using foraging theory models. *World Archaeology* 34: 84–102.

Nagaoka, Lisa. 2005. Differential recovery of Pacific Island fish remains. *Journal of Archaeological Science* 32(6): 941–955.

O'Connell, James F., Jim Allen, Martin A. J. Williams, Alan N. Williams, Chris S. M. Turney, Nigel A. Spooner, Johan Kamminga, Graham Brown, and Alan Cooper. 2018. When did *Homo sapiens* first reach Southeast Asia and Sahul? *Proceedings of the National Academy of Sciences* 115(34): 8482–8490.

Olsen, Bjonar, Michael Shanks, Timothy Webmoor, and Christoper Witmore. 2012. *Archaeology: The Discipline of Things.* Berkley, CA: University of California Press.

Orton, David. 2010. Both subject and object: Herding, inalienability and sentient property in prehistory. *World Archaeology* 42: 188–200.

Overton, Nicholas J., and Yannis Hamilakis. 2013. A manifesto for a social zooarchaeology. Swans and other beings in the Mesolithic. *Archaeological dialogues* 20(2): 111–136.

Pillay, Patricia, Melinda S. Allen, and Judith Littleton. 2022. Canine companions or competitors? A multi-proxy analysis of dog-human competition. *Journal of Archaeological Science* 139:105556.

Reitz, Elizabeth J., and Elizabeth S. Wing. 2008. *Zooarchaeology* (2nd ed.). Cambridge: Cambridge University Press.

Richards, Michael P., Eric West, Barry Rolett, and Keith Dobney. 2009. Isotope analysis of human and animal diets from the Hanamiai archaeological site (French Polynesia). *Archaeology in Oceania* 44(1): 29–37.

Ritvo, Hariett. 2007. On the animal turn. *Daedalus* 136(4): 118–122.

Russell, Nerissa. 2012. *Social zooarchaeology: Humans and animals in prehistory.* Cambridge: Cambridge University Press.

Sand, Christophe. 2000. The Specificities of the "Southern Lapita Province": The New Caledonian Case. *Archaeology in Oceania* 35: 20–33.

Serpell, James. 1995. From paragon to pariah: Some reflections on human attitudes to dogs. In *The domestic dog: Its evolution, behaviour and interactions with people.* James Serpell, ed. Pp. 245–256. Cambridge: Cambridge University Press.

Serpell, James A. 2009. Having our dogs and eating them too: Why animals are a social issue. *Journal of Social Issues* 65(3): 633–644.

Shannon, Laura M., Ryan H. Boyko, Marta Castelhano, Elizabeth Corey, Jessica J. Hayward, Corin McLean, Michelle E. White, Mounir Abi Said, Baddley A. Anita, and Nono Ikombe Bondjengo. 2015. Genetic structure in village dogs reveals a Central Asian domestication origin. *Proceedings of the National Academy of Sciences* 112(44): 13639–13644.

Shaw, Ben J., Glenn R. Summerhayes, Hallie R. Buckley, and Joel A. Baker. 2009. The use of strontium isotopes as an indicator of migration in human and pig Lapita populations in the Bismarck Archipelago, Papua New Guinea. *Journal of Archaeological Science* 36(4): 1079–1091.

Smith, Anita. 2002. *An archaeology of West Polynesian prehistory.* Canberra: Pandanus Books.

Specht, Jim, Tim Denham, James Goff, and John Edward Terrell. 2014. Deconstructing the Lapita Cultural Complex in the Bismarck Archipelago. *Journal of Archaeological Research* 22(2): 89–140.

Spennemann, Dirk. 1990. The role of pigs and dogs in the taphonomy of modern and archaeological faunal assemblages from Tonga. In *Problem Solving in Taphonomy: Archaeological and palaeontological studies from Europe, Africa and Oceania.* Su Solomon, Iain Davidson, and Di Watson, eds. Pp. 101–107. St. Lucia, QLD: Anthropology Museum, University of Queensland.

Spriggs, Matthew. 1996. What is Southeast Asian about Lapita? In *Prehistoric Mongoloid Dispersals.* T. Akazawa and E. Szathmary, eds. Pp.322–346. Oxford: Oxford University Press.

Steward, Helen. 2009. Animal agency. *Inquiry* 52: 217–231.

Storey, Alice A., Andrew C. Clarke, Thegn Ladefoged, Judith Robins, and Elizabeth Matisoo-Smith. 2013. DNA and Pacific Commensal Models: Applications, Construction, Limitations, and Future Prospects. *The Journal of Island and Coastal Archaeology* 8(1): 37–65.

Sutton, Alice, Mary-Jane Mountain, Ken Aplin, Susan Bulmer, and Tim Denham. 2009. Archaeozoological records for the highlands of New Guinea: A review of current evidence. *Australian Archaeology* 69(1): 41–58.

Sykes, Naomi. 2014. *Beastly questions: Animal answers to archaeological issues.* London: Bloomsbury Publishing.

Titcomb, Margaret. 1969. *Dog and man in the ancient Pacific, with special attention to Hawaii.* Honolulu: Bernice P. Bishop Museum.

Tsang, Roxanne, Sebastien Katuk, Sally K. May, Paul S. C. Taçon, François-Xavier Ricaut, and Matthew G. Leavesley. 2022. Rock art and (re)production of narratives: A cassowary bone dagger stencil perspective from Auwim, East Sepik, Papua New Guinea. *Cambridge Archaeological Journal* 32(4): 547–565.

Watts, Vanessa. 2013. Indigenous place-thought and agency amongst humans and non humans (First Woman and Sky Woman go on a European world tour!). *Decolonization: Indigeneity, Education & Society* 2(1): 20–34.

Wickler, Stephen. 2004. Modelling colonization and migration in Micronesia from a zooarchaeological perspective. In *Colonisation, migration, and marginal areas: A zooarchaeological approach*. Mariana Mondini, Sebastian Munoz, and Stephen Wickler, eds. Pp. 28–40. Oxford: Oxbow Books.

Wilmshurst, Janet M., Terry L. Hunt, Carl P. Lipo, and Atholl J. Anderson. 2011. High-precision radiocarbon dating shows recent and rapid initial human colonization of East Polynesia. *Proceedings of the National Academy of Sciences* 108 (5): 1815–1820.

Wood, Jamie R., Andrea Crown, Theresa L. Cole, and Janet M. Wilmshurst. 2016. Microscopic and ancient DNA profiling of Polynesian dog (kurī) coprolites from northern New Zealand. *Journal of Archaeological Science: Reports* 6: 496–505.

Wronska-Friend, Maria. 2015. From shells to ceramic: Colonial replicas of Indigenous valuables. *Journal of Museum Ethnography* 28: 50–69.

Zhang, Zhe, Saber Khederzadeh, and Yan Li. 2020a. Deciphering the puzzles of dog domestication. *Zoological Research* 41(2):97.

Zhang, Ming, Guoping Sun, Lele Ren, Haibing Yuan, Guanghui Dong, Lizhao Zhang, Feng Liu, Peng Cao, Albert Min-Shan Ko, and Melinda A. Yang. 2020b. Ancient DNA evidence from China reveals the expansion of Pacific dogs. *Molecular Biology and Evolution* 37(5): 1462–1469.

Živaljević, I. 2021. Multispecies pasts and the possibilities of multispecies futures in the age of the Anthropocene. *Етнёантрёпёлёшки прёблеми* 16(3): 659–676.

6 Dog or Dog-Gone? A View from Four (Almost) Dogless Islands

Richard Feinberg and Cathleen Pyrek

Introduction

Since 1972, Rick Feinberg has conducted research with three small Polynesian "outlier" communities: Anuta and Taumako Islands in the Solomons' Temotu Province and Nukumanu Atoll in Papua New Guinea's Bougainville Autonomous Region. In addition, Feinberg and co-author, Cathy Pyrek, worked together in 2012 on Atafu Atoll in Tokelau. There were no dogs on either Anuta or Nukumanu during the times that Feinberg was there, nor were they, in his recollection, a significant topic of conversation among islanders in either locale. After reviewing thousands of pages of field notes, Feinberg's memory was confirmed: he found virtually no references to canine residents on either Anuta or Nukumanu. For Taumako such references were not absent but sparse. However, travel to or from small, remote outposts typically requires one to spend time on larger islands. In the Solomons, those are Guadalcanal and Ndeni, sites of the national and provincial capital respectively. Travel to Nukumanu required stays in Port Moresby on the main island of New Guinea as well as Arawa on Bougainville or, later, Buka, the site of provincial governance following a secessionist movement on Bougainville in the 1990s. Similarly, travel to Tokelau typically involves spending time in Āpia, Samoa's capital. The larger islands provided an opportunity to observe these islanders' attitudes toward and interactions with dogs.

Our observations have ecological implications. On small islands with limited carrying capacity, dogs compete with people for scarce resources. Therefore, human inhabitants are likely to conclude that the risks of raising canines outweigh potential rewards, and the number of dogs is small to non-existent. On larger, more resource-rich islands, the calculus changes, and keeping dogs holds more appeal. This ecological distinction has been noted elsewhere (Cramb 2022; Koungoulos, this volume, Chapter 7). However, as residents of small islands with little dog-related experience emigrate to larger landmasses, interactions are often awkward and uncomfortable for dogs and humans alike. Over time, that relationship may change to some extent. Nonetheless, the process is likely to be fraught and the outcome uncertain.

DOI: 10.4324/9781003658696-6

The latter point carries implications for what contributors to this volume, following Lohmann (this volume, Chapter 1), term "human–canine interspecies culture". For present purposes, we accept the expression, but with some qualifications. One of us came of age, anthropologically, with the idea that culture is a system of symbols and meanings that are shared by members of a community and transmitted across generations (see, e.g., Parsons and Shils 1962; Schneider 1968; Feinberg 1979). Since E. B. Tylor (1871) famously introduced the term into anthropology, "culture" has been defined in literally hundreds of ways (e.g., Kroeber 1952; Kroeber and Parsons 1958; Kroeber and Kluckhohn 1963; Feinberg 1979). Definitions are not intrinsically right or wrong; rather, they are more or less useful. For ethnographic and analytical purposes, we have generally found the narrower definition, distinguishing culture from the social system but recognizing that the two are interrelated, to be most productive.

Specialists in animal behavior often use a broader definition. Thus, Whiten (2021a, 2021b), in discussing "animal culture", portrays it as "the inheritance of an array of behavioral traditions through social learning from others" (2021a:1) or "all that is learned from others and is repeatedly transmitted in this way, forming traditions that may be inherited by successive generations" (2021b:1). Similarly, Guillo and Claidière (2020:1) define culture as "the result of direct social transmission from a model to a conspecific". In that sense, dogs (as well as many other animals) unquestionably have culture. Furthermore, it clearly is the product of interaction both with other dogs and members of divergent species, including human beings.

Parsonians, including Geertz, Schneider, and their intellectual progeny, are likely to describe such characteristics as patterns not of culture but of social action. And to those using the narrower definition, evidence that dogs share systems of symbols and meanings among themselves, or that they pass such a system from one dog to another, remains questionable. Thus, Carucci (this volume, Chapter 9: 180) argues that dogs, unlike humans, do not "possess cultural abilities grounded in the manipulation of symbols". Similarly, Dobrin (this volume, Chapter 2: 55) notes that, among the Arapesh, dogs and humans have "no shared language, [so] speaking must take place through signs and gestures". At the same time, both agree—and we concur—that effective interspecies communication occurs; it takes different forms in different communities; and the result can be striking contrasts in behavior patterns. As Lohmann (this volume, Chapter 1: 16) notes, in societies that include dogs as well as humans, "not just dogs, but their humans as well undergo changes in their mental, behavioral, health, developmental, genetic, and artifactual expressions of culture that would not exist without both species participating". As socially learned information that requires the participation of multiple species, the idea of "interspecies culture" can be usefully employed.

Dogs and humans interact in multiple ways, learning from one another and adapting their behavior. Canines in different communities manifest different behavior patterns. People around them behave in different ways, thereby either reinforcing their non-human companions' habits and expectations or leading them to modify those habits. Then, as dogs' behavior changes, that of their human acquaintances adapts as well. All of this is illustrated by our observations on four small islands and their larger counterparts.

Anuta

Anuta is the easternmost populated island in the Solomons. It is approximately one kilometer in diameter, rises to an altitude of 70 meters, has a wide array of ecological zones, and is surrounded by reefs extending around five km beyond the seashore.[1] When Feinberg was first there, in 1972–1973, the resident population was approximately 150 people. At the time of his most recent visit, in 2013, the population had grown to almost 300. In addition, several dozen Anutans usually live in other parts of the Solomons.[2]

The Anutan word for dog, *korii*, is widespread in Polynesia, suggesting a connection sometime in the distant past. However, during Feinberg's visits to Anuta in 1972–1973, 1983, 2000, 2007, and 2013, no dogs were present on the island. When he asked why this was the case, islanders explained simply that they had limited land resources. For similar reasons, Anutans do not keep swine. On a few occasions, islanders tried raising pigs for food, but they created more problems than they were worth, and after a brief period of experimentation they were permanently banned.[3]

As of 1972, virtually all Anutan men and a few Anutan women had spent a year or more living overseas, either in the colonial (now the national) capital of Honiara or working on the Levers copra plantation in the Russell Islands.[4] This trend began in the 1950s, accelerated through the 1960s, and has continued ever since. Except for a pause during the civil war (known euphemistically as "The Tension") that shook the Solomons around 2000, when many outer islanders returned home to escape the fighting, the number of Anutans in the central Solomons has continued to expand. Some work for wages; others have attended schools on several islands. In their travels, Anutans have encountered dogs and developed some sense of their potential value as well as the costs. They knew of people keeping dogs for security purposes but were also aware of canines attacking innocent people who were simply going about their business.

On Anuta, there was relatively little concern about security, so the main reason to bring dogs to the island would have been for companionship. Such companionship seemed appealing to a few islanders, but any such value was nullified by the risk of attack. Moreover, Anutans reported that while they raised sufficient food to sustain the human population, they lacked the surplus needed to feed dogs. They had access to plenty of fish, and when fish were unavailable, they hunted birds. In addition, they raised chickens which they sometimes slaughtered and consumed as a source of animal protein. Anutans were generally well fed, but they were aware of their food supply's potential vulnerability. Indeed, Feinberg's first visit in 1972 came months after a devastating hurricane that had destroyed the island's coconut, papaya, and other staple crops.

Being aware of potential food insecurity, Anutans bury starchy vegetables and fruits in underground pits, where they ferment and are preserved for years.[5] Despite the availability of fermented vegetables, called *ma*, however, the islanders' diet was substantially curtailed, and the senior chief imposed strict regulations to ensure that no one would starve.[6] While Islanders understood that dogs are primarily carnivorous, they were aware that canines consume vegetables as well. More importantly, perhaps, there was concern that dogs might attack chickens and wild birds, raid the stores of fish that were being preserved for future meals, and perhaps kill cats. Islanders kept a large number of cats, which were valued for control of the rat population. Cats were recognized as more effective than dogs for that purpose, so the priority was on protecting cats (cf. Carucci, this volume, Chapter 9).

From the Anutans' viewpoint, rats were a nuisance. They never ate rats, and the rodents would maraud their gardens as well as packages of stored food. They also would gnaw holes in cloth and deposit feces. Anutans' cats, then, served a clear purpose. Occasionally, people shared food with their cats. For the most part, however, felines either foraged for themselves or ate leftovers that their human companions had discarded. By contrast, dogs were depicted as competing with humans for limited staples. They were perceived to serve no practical purpose on the island, and Anutans' limited experience with dogs during their visits to larger isles in the Solomons did little to instill a sense that they might provide emotional support to their human counterparts.

While Feinberg was on Anuta, neighbors told him of an islander who once attempted to keep a dog. The dog proved, however, to be more of a nuisance than an asset, reportedly attacking and sometimes killing chickens. He also threatened people and occasionally bit them, so islanders decided to get rid of him.

Mark Rongokavea, a young Anutan currently living in Honiara, tells a more elaborate story. He reported to Feinberg via e-mail (personal communication) that there were a few additional attempts to keep dogs on Anuta. According to Mark, a total of four dogs at different times have been brought to the island, and they all died of natural causes. He added, "yes, the dogs did cause some problems such as killing chickens and cats, but [to his knowledge] no one was bitten by any of the dogs." Two of them were tied full-time, and only one was able to run freely. Two of the dogs were brought from the neighboring island of Tikopia. One accompanied a Tikopian primary school teacher who stayed on Anuta for a few years. Another was supplied by an Anutan who lived near Honiara. All four died before Feinberg's arrival in 1972.

Mark confirmed Feinberg's impression that Anuta's experiment with keeping dogs was short lived, and no one has suggested that it be repeated. In some regions, dogs have been habituated to exercise self-discipline, refrain from threatening humans, and live peacefully with other animals.[7] Such canine behavior, however, would be atypical in the experience of most Solomon Islanders.

The picture among Anutans who have migrated to other islands of the Solomons is different from that on the home island. Most of those migrants have settled in the Honiara area, on the large island of Guadalcanal. They are engaged in a variety of occupations and, for the most part, interact cordially with other Solomon Islanders. Typically, however, they identify with their place of origin and have closest relations with members of their home community (*te kanopenua*). A number of Anutan households are scattered through the Honiara region, but the major concentration is in White River, a settlement just west of the city limits. Feinberg has spent time in White River on numerous occasions, most recently in 2013, and he found canine encounters to be common. Still, he was surprised to see Mark's account of Anutans themselves keeping dogs.

Most unexpected was the story of one man who lived just outside of the main Anutan enclave and reportedly kept a large number of dogs: perhaps ten or more. By the late 1990s, according to Mark, his home was "surrounded by his dogs".

Mark also stated that in the mid-1990s, when he was a young child, his family lived in Mamara, a settlement near Honiara, and kept two dogs. He described the dogs as well behaved but was uncertain of how other Anutans perceived them. Solomon Islanders, according to Mark, keep dogs both as pets and for security, a point he vividly depicted in a detailed e-mail.

One of the family's dogs, named Blackie, was an aggressive dog and would break the chains that restrained him. The other, Snowie, was a docile, obedient companion. They kept Blackie for security purposes after an incident where strangers threatened the family. Over the years of keeping these two dogs, Blackie remained aggressive and disobedient while Snowie demonstrated his loyalty on numerous occasions.

In Mark's communication, he discusses several themes. Blackie's aggressiveness illustrates the value of dogs for security purposes. In classic fashion, he—then a young child—and a dog began their relationship as mutually suspicious, learned to trust one another, bonded, and eventually developed an emotional attachment. Mark found that regular feeding played an important role in that development of trust.

The contrast between Blackie and Snowie illustrates the differences in individual personalities among dogs who may live in the same household but have experienced strongly contrasting relations with their humans. As time went on, Blackie continued to be somewhat guarded, while Snowie consistently went out of his way to cooperate with his human companions. This was particularly impressive in Snowie's display of self-discipline when, on one occasion, he was provided the opportunity to help himself to a cooked fish but had been admonished that such self-indulgence was against the rules. He refrained from consuming the fish until Mark's mother returned and relieved him of his duties.

Inter-species concern and affection is evident in a description of the two dogs waiting for Mark's family to return home from White River, their obvious excitement when their people finally arrived, and Snowie growing sufficiently concerned to venture out and look for his people one evening when they were alarmingly late. Such a relationship is unusual in our experience with dogs and people in the western Pacific. It shows both the variability of individuals within a larger community and the potential to develop close interspecies relationships, even in a setting where interactions between dogs and humans are often tense.

Guadalcanal, site of the Solomon Islands capital, has over 150,000 residents from diverse ethnic and linguistic groups, many of which see one another as rivals. The Honiara area also has the country's greatest concentration of businesses and educational institutions. Social stratification and crime are facts of life, and many residents keep dogs for security purposes. Owners may feed the dogs enough to keep them from running off. For many dogs, however, a large portion of their diet comes from trash left on the streets. They are often fearful of approaching humans and retreat when one approaches. At night, or when they gather in packs, they may become aggressive. A person coming close to someone else's dog is warned with barks and growls to back away, and anyone who fails to heed the warning risks being bitten. That reality contrasts sharply with the experience described by Mark, where the dog is well cared for. This illustrates the point that different dogs and different people may have dramatically variable relationships, even within one relatively small country. Perhaps it suggests that the idea of "multiple voices"—i.e., different understandings of reality and behavioral norms within a single community (see, e.g., Clifford 1988; Clifford and Marcus 1986; Feinberg 1994; Feinberg and Zimmer Tamakoshi 1995)—may apply to dogs as well as humans.

Sometimes dog packs roam the streets and threaten people. Common understanding is that, should one encounter an aggressive dog, the most effective deterrent is to attack the animal by throwing stones. Dogs have learned the lesson, so typically one only need bend down as if to pick up a stone, and the dog will retreat (cf. Lohmann, this volume, Chapter 8; Prokosch, this volume, Chapter 3). Nonetheless, mild-mannered humans, or those less physically adept, often feel intimidated by dogs. That is particularly true of people coming from small islands, who have no experience interacting with canines at home. Their impressions are grounded in their observations of dogs on Guadalcanal and their interaction with others as well as their own direct experience.

While Anutans do not keep dogs or pigs on their home island, they do occasionally keep frigate birds as pets. They let the birds fly free, but people feed them, and they typically return at the end of the day to their human companions. Feinberg also recalls one Anutan living in Rove, on the western edge of Honiara, keeping a parrot. That was in the 1980s.

Anutans on the home island raise chickens. They often feed them food scraps such as leftovers from meals or shredded coconut meat after the cream has been squeezed out. The chickens run free, with little interaction between them and their human overseers—other than when they are slaughtered for food. In addition, when Feinberg was first on Anuta, community members had two ducks, whom they treated in the same way as the chickens. A few weeks into his stay, the ducks were eaten and were never replaced.

In summary, Anutans exhibit varied attitudes toward different animals. They detest rats, which are dirty, leave feces, and raid the food on which humans rely. Cats are appreciated for helping keep the rat population under control. They also largely forage for themselves or are fed leftovers from people's meals. Chickens are appreciated as a source of protein, used mostly when fish and wild birds are unavailable. Frigate birds are mostly kept for their aesthetic value. They hunt for themselves at sea and do not compete with humans for food. Lastly, pigs are recognized as a potential high-status protein source that might be used at major feasts. However, they damage the environment and compete with humans for food. In that respect, dogs most closely resemble pigs, but with the further liability that Anutans do not eat them. In short, Anutans see dogs as more risk than reward and have little interest in keeping them on the home island.

Nukumanu

Nukumanu Atoll, like Anuta, is a small, remote Polynesian outpost. It is the easternmost community in Papua New Guinea, close to the Solomon Islands border, and only about 40 km north of Ontong Java Atoll in the Solomons' Malaita Province. In 1984, Feinberg was there for three months and spent two more months with Nukumanu expatriates in other parts of Papua New Guinea. In 2000, as Bougainville's war of secession was drawing to a close, he spent an additional three months with the Nukumanu community in the nation's capital, Port Moresby, and on Buka Island in what was then the Bougainville Autonomous Region.

As an atoll, Nukumanu's marine resources are abundant. However, it has porous soil, and its rainfall and retention of fresh water are limited, making vegetation relatively sparse. Despite the atoll's limited land resources, the Nukumanu raised a few pigs. However, like the Anutans, they did not keep dogs.

As is true of Anutans, many Nukumanu have moved away from their home island. In 1984, a number had congregated on Bougainville, where the Panguna Mine provided a major source of employment. In the 1990s, what had been North Solomons Province was torn apart by military conflict. The gold-and-copper mine in Bougainville's Crown Prince Mountain Range, and the administrative center in Arawa on the island's eastern coast, were shut down. Most Nukumanu left. By the time of Feinberg's return in 2000, the fighting had largely subsided, and most Nukumanu had moved either to the new administrative center on Buka—a much smaller but still sizeable island just north of Bougainville—or to the national capital, Port Moresby. Feinberg spent time with the Nukumanu communities both in Moresby and on Buka, and he was unaware of anyone keeping dogs in either locale. Both communities were small, each hosting a few dozen people. In both cases, once one left the Nukumanu settlements, dogs might be encountered. Their treatment was similar to what Feinberg had observed in the Solomons, and Nukumanu responses to local dogs resembled those of most Anutans living on Guadalcanal: the costs associated with their presence outweighed the benefits; therefore, by and large, they were not welcome.

Taumako

Taumako, in many ways, resembles Anuta. It is a small Polynesian Island in Temotu Province of the southeastern Solomons. Like Anuta it is of volcanic origin, but it is larger: approximately five km long, three km wide, and rising to an altitude of 300 meters. In contrast with Nukumanu, Taumako's terrestrial landscape hosts a wide variety of ecological zones. Its people also claim control of several smaller offshore islands. In proportion to Taumako's overall size, it has less flat ground than Anuta for cultivating taro and comparable crops, but islanders tend ample gardens on the mountainside.

Archaeological evidence suggests that atolls and low islands are likely to experience repeated introductions and loss of dogs (Cramb 2022). By contrast, dog populations tend to remain once introduced on larger, high islands. Taumako underscores this point both archaeologically and from the viewpoint of contemporary ethnographic research.

In absolute terms, Taumako is a small island. However, its maximum altitude is more than four times that of Anuta. And, while atolls may be spread over hundreds of square kilometers of ocean, their land mass and terrestrial terrain are dwarfed by Taumako's. The Taumako take advantage of their island's size and varied terrain to raise enough pigs that they are common fare at major feasts. And, of the four islands on which this chapter focuses—Anuta, Nukumanu, Taumako, and Atafu in Tokelau—Taumako, with the largest land area and most rugged landscape, is the only one where dogs are found.[8]

Archaeologists Foss Leach and Janet Davidson (2008:26) report that, as of 1977, "a considerable population of wild dogs" lived on the Taumako group's outer islands of Lakao and Tuleki. They add:

> A dog skull with extreme wear on all teeth was found on the surface at Ako, and another was on the surface at Aua .... There are significant numbers of wild dogs on Lakao and they can be heard howling in the distance at night. During our excavations at Tavatava, some visited our camp at night looking for food. The workmen set special dog traps and caught one emaciated bitch, which was dispatched. There is little useful food for carnivores on these islands except land crabs and small lizards and these dogs use their teeth to open coconuts for the meat. No doubt this accounts for the extreme tooth wear .... As we left Tuleki Island further north, on a later occasion, a dog came out of the bush and swam out to sea attempting to get into our canoe, presumably to escape from an island with so little food
>
> (Leach and Davidson 2008:41).

In addition to encounters with live dogs, archaeological excavations reported in Leach and Davidson's volume on *The Archaeology of Taumako* suggest a long-time, albeit less-than-common, canine presence:

> Only two dogs were identified from this site, one from the Crust of Layer 5 in Area C, and the other from Layer 3B in Area A. No dog remains were found in the main village deposits. Dogs were present, although rare, in the more recent site at Kahula on Taumako (See Chapter 7). We can therefore be certain that dogs were present in the Taumako group from the period of Level I at Te Ana Tavatava 900 BC to 400 BC. However, it is not possible to determine, from existing evidence, whether dogs were present on Taumako throughout the sequence, or whether they died out and were later reintroduced, perhaps by newly arrived Polynesian speakers. So few dog bones were found in dated archaeological provenances on Taumako that little can be said about their biological affinities. However, Smith (Appendix 3) was able to comment on modern Taumako dogs, whose skeletal remains were found on Lakao.[9] He concludes that these are similar in a number of respects to prehistoric Polynesian dogs. They have long narrow crania, high cranio-facial indices, and lack of tooth crowding. In this respect, they are similar to New Zealand prehistoric dogs. However, New Zealand dogs have prominent sagittal crests, indicating very strong jaw muscles. The Taumako crania are more gracile. The single post-cranial skeleton has short legs in relation to body size, again comparable to New Zealand prehistoric dogs
>
> (Leach and Davidson 2008:90–91).

(See also Titcomb and Pukui 1969; Koungoulos, this volume, Chapter 7).

Feinberg spent almost a year with the Taumako community in 2007–2008, most of that on Taumako Island itself. During his time on the island, he did not find that dogs (*kulī*) were common. Nonetheless, they were a fact of life.

A few families kept them as watchdogs and pets. None of those were households with whom Feinberg stayed and, since he did not live with dog owners, he had limited opportunity to observe dog–human interaction. However, people did discuss canines, and those conversations occasionally appear in his field notes.

In Feinberg's experience, Taumako dog owners did not keep pets inside their houses. That is not surprising, as people typically lived in small leaf houses with just one or two rooms. Theoretically, islanders could have shared their sleeping mats with pet dogs, but Feinberg saw no evidence of such a close relationship. For the most part, dogs were left to forage for themselves or were fed leftovers from their owners' meals. Similarly, cats roamed pretty much at will. Chickens were allowed to run about and were not penned. Pigs, by contrast, were tied up outdoors: sometimes near the owner's house; more often in a fairly distant garden.

Taumako speak of dogs having connections to human families that English speakers might describe as "ownership". The numbers involved, however, are small, and Feinberg never heard people on the island allude to trading or selling dogs. Occasionally, dogs left their human companions, became feral, and caused problems by eating chickens and attacking pigs. Despite the island's small size, the dense forests and steep hillsides provide feral dogs with room to hide. Therefore, attacks on other animals posed an ongoing problem.

As of 2008, people identified four dogs that had been attacking their pigs. Feinberg's neighbor, Basil Tavake, killed one with a spear and teamed up with his brother, Basil Mekau, to try to catch and kill a second. The latter quest was unsuccessful. There were reports of someone on the far side of the island having killed a third wild dog, but at least two remained and continued to cause havoc.

An additional preoccupation with dogs among Taumako centered on the spirit world (see Feinberg 2019; cf. Van der Ryn, this volume, Chapter 10). According to traditional convictions, which most Taumako still embrace, each large-scale descent group (*kaenga*) is guarded by a protective spirit which might be described as a totem. Such spirits are called *sua te kaenga*, and they help protect their people from disaster. Often, they serve as omens, informing humans of events that have occurred outside their fields of vision or warning them of forthcoming illness, death, and other calamities. One *kaenga* is called Matalele, and its totem is a dog. This is thought to be a particular dog (Te Kulī) rather than a generic class. Occasionally Taumako report having spotted Matalele's totem, whom they are able to differentiate from ordinary dogs because of his distinctive appearance. Matalele's spirit dog is said to be large and white, with long legs and a white and black face.

Dogs hold a special place in Taumako culture, but that place is in the spiritual more than the physical world. Taumako have recent memory of dogs presenting more risk than reward and, overall, do not value their presence.

Tokelau

In 2012, the authors spent a month together on Atafu Atoll in Tokelau, a group of three atolls located north of Samoa and politically affiliated with New Zealand. We spent another month with the Tokelau community in Āpia, Samoa.

As an atoll, Atafu is much like Nukumanu. With sandy, coralline soil, low elevation, and limited rainfall, it is difficult to raise most crops. Marine resources, however, are abundant. Because of Tokelau's connection to New Zealand, its material infrastructure is more highly developed than Nukumanu's. Ships are likely to visit every week or two, as opposed to a few times per year. People live in concrete multi-room—and often multi-story—houses with electricity and indoor plumbing. They hunt birds and coconut crabs, and they raise pigs. However, as on Nukumanu and Anuta, which are—like Atafu—small, remote outcroppings with limited resources, dogs are wholly absent.

Atafu has not always been dog free. Archaeological research performed by Simon Best (1988) found evidence of dogs dating back approximately 1000 years. And our friend, Ahela Huka, pointed out to us that one of Atafu's islets is named Tangi a Kulī. *Kulī* or *kurī* is a common Polynesian word for "dog" (cf. Anutan *korii* and Taumako *kulī*), and *tangi* means "to cry" or "to cry out". Ahela translated Tangi a Kurī as "Barking of the Dog".

When Ahela was a young child, living on Atafu, his grandmother kept a dog named Lova. However, Lova started biting children, and the community government ordered the dog to be put down. From that time onward, dogs have been banned on the atoll.

During our time in Tokelau, we encountered a few families who "kept" or looked after non-human animals. As on Anuta, these were animals that competed but minimally, if at all, with people for resources. One family had a pet albatross as well as turtles. From Pyrek's field notes, the family of Malia, Ahela's cousin:

> has a pet albatross named Bob. He's very tame and will allow strangers to pet him. They also have two turtles. Several families have pet turtles because apparently someone managed to harvest some eggs once and gave baby turtles to people all over the island. Unfortunately, everyone keeps their turtles in large plastic bins, which seems a miserable existence.

Bob the albatross, unlike the turtles, was not confined and spent much of his time parked on Malia's boat, which was docked in the lagoon. He could be counted on to join the group during outdoor cooking activities.

Travel to and from Tokelau involves catching a ship in Āpia. Samoa, like Tokelau, has long had a close relationship with New Zealand. Consequently, one finds Tokelauan expatriate communities both in New

Zealand and Samoa, and Tokelauans have lived in Āpia dating back at least as far as 1852 (Huntsman 2004). As is true of Anutans and Taumako in Honiara, or Nukumanu in Port Moresby and Buka, Tokelauans in Āpia frequently encounter dogs. These may be owned by local Samoans or be strays, wandering the city's streets. With a few exceptions, Tokelauans themselves do not keep dogs.

Our stay in Āpia coincided with the 50th Anniversary of Samoan Independence. This fact proved to be significant with regard to the dog population. Upon arriving in Āpia, we were immediately aware of the presence of many dogs. A few were owned by humans, but a large number appeared to be feral. They roamed the streets, occasionally threatening us as we walked. We could always scare them off with a few small rocks thrown in their direction (cf. Prokosch, this volume, Chapter 3; Carucci, this volume, Chapter 9). As Roger Lohmann (personal communication, 2023) has noted, this creates a circular pattern in which "dogs and humans threaten one another and train one another to be fearful and threatening toward one another". Dogs in Samoa also seemed remarkably adept at avoiding vehicular traffic, moving to the road's edge as cars approached. This may be an example of "interspecies culture". It appears that avoiding traffic took root to the point that humans scarcely worry about hitting dogs with their cars. We were impressed by the presence and behavior of these dogs.

Then, suddenly, just before the Independence Day festivities, all the dogs were gone. Because it was the 50th anniversary, dignitaries and expatriates came from countries far and wide to celebrate. Mysteriously, no feral dogs were anywhere to be found; they literally all vanished overnight. We asked about them and no one seemed to know what we were talking about. We were never able to affirm what happened but suspect that they were rounded up and euthanized.

During our stay in Āpia, we lived with expats from Atafu in a section of town known as Lotopā. Our hosts were among the few Tokelauans who kept a dog. She was named Bingo and was less a pet than a working watchdog. Samoa's rate of violent crime is low compared to such US enclaves as the South Side of Chicago, or Papua New Guinea's Port Moresby. Still residents of Āpia are concerned and, when possible, take precautions. Pyrek was struck while in Samoa and Tokelau by how frequently the concept of danger "over there" arose. Whenever she discussed where she was from, for instance, the first point people raised was the "danger". To our knowledge, none of her interlocutors had been to the Americas. Despite their lack of personal familiarity with the US, however, they appeared convinced that it is an exceptionally dangerous place.[10] This persistent concern about danger is likely the reason they kept Bingo, who was fed table scraps. In return, she stayed with her humans and kept watch. She did not allow humans to get near her, and no affection was shown in either direction.

Bingo was pregnant when we first met her. She had her pups while we were on Atafu. When we returned to Lotopā, one of her pups had won the heart of the man who was head of the household, Lafu Solomona. He held the puppy and played with it, laughing and enjoying the company. It was a remarkable change from the way we had seen people interacting with dogs. Lafu pondered what to name the puppy. Pyrek suggested the name of her childhood dog, Penny. Lafu agreed that Peni was a good name for a dog, and that is how she came to be known to the household. This episode reinforces our observations concerning Anuta's Mark Rongokavea, his relationship with the dog, Snowie, and the possibility of developing interspecies affection even in settings where such connections tend to be discouraged.

General Observations

Dogs are uncommon or entirely absent on the small islands where we have worked. That is unsurprising, as they require expenditure of resources that could otherwise be used to support the human population. This is also true of cats, which are more common in such surroundings. However, cats consume less food and are appreciated for keeping rats in check. Dogs bother cats and pigs, are at least as likely as cats to molest chickens and wild birds, and are less efficient than cats as rat hunters. As seen in other regions, with adequate care and training, dogs can be taught to behave differently. Under the right conditions, they might learn to live in harmony with chickens, cats, and pigs while serving as effective assets for control of rats. In our Pacific travels, however, and despite the emergence of interspecies cultural adaptations, we have not observed attempts at such training (cf. Prokosch, this volume, Chapter 3).

People from the islands we have visited occasionally care for animals other than dogs. Anutans enjoy the company of frigate birds; and on Atafu, turtles and an albatross named Bob were part of daily life. As Carucci (this volume, Chapter 9) observes, pets may have more than purely functional value. And, at least with birds, there seems to be reciprocal appreciation: the birds are free to leave; yet, they stay close to humans who enjoy their company.

When people move to larger islands, some interaction with dogs is unavoidable. Migrants from small islands often fear dogs, largely because they lack experience in dealing with them. Some learn to appreciate dogs for their ability to warn of unwelcome intruders and discourage crime. However, even those who value dogs' positive functions rarely treat them with the love and warmth that Westerners often invest in pets (cf. Prokosch, this volume, Chapter 3). Our observations in urban areas of larger islands correspond with those of other contributors to this volume.

As among New Guineans and Marshall Islanders (e.g., Carucci, this volume, Chapter 9; Prokosch, this volume, Chapter 3), dogs in our fieldwork experience are rarely treated empathetically, much less indulged. Strays avoid human interaction when possible and, when offered no alternative, they may become aggressive. Occasionally, dogs are valued as household members and treated as pets. That, however, is unusual and, even in such cases, they tend not to live inside their owners' homes. The only reference to feral dogs we have encountered among the smaller islands is on Taumako, where they are perceived to be a threat and actively hunted in an effort to protect other animals.

The fraught relationship that people have with dogs makes sense when measuring pragmatic costs and benefits. Nonetheless, Mark Rongokavea's relationship with his family's dogs, and Lafu Solomona's affection for the puppy, Peni, make clear that islanders, at times, treat dogs with warmth and care. Such positive connections, although exceptional, may help explain a canine presence in the archaeological record as well as human life today.

Notes

1 For a discussion of Anutan agriculture and the island's terrestrial zones, see Yen 1973. For the island's seascape and mental mapping of the ocean floor, see Feinberg et al. 2003.
2 For an overview of Anuta and its social organization, see Feinberg 2011[2004].
3 Shortly before Feinberg left Anuta in 1973, one family imported a couple of piglets. Children in the family felt the pigs were cute and declared affection for them. However, by the time of Feinberg's next visit, in 1983, they had been long gone and no one expressed an interest in repeating the experiment.
4 These movements are detailed in the appendices of Feinberg 1981. For more on the Levers' plantation and the Polynesians working there, see (Larson 1966, 1968).
5 Some other Polynesians in times past have followed similar procedures as a hedge against famine. With the exception of Anuta, however, the practice appears to have been abandoned.
6 Although no one was in immediate danger of starvation, it was unclear how severe the famine would become. Thus, the senior chief declared that the entire island would behave as a single domestic unit, and crops from every garden would be treated as belonging collectively to the entire community. As the chief put the matter, he wanted to be sure that if anyone were to die of starvation, "we all will die together" (see Feinberg 2011:180–181).
7 For example, in 2019 Feinberg spent a semester in the Czech Republic. He had a bicycle which he used as his primary means of local transportation, and he was amazed that dogs he passed on the street or sidewalk invariably ignored him. Under comparable circumstances in the US, even calm and non-aggressive dogs are likely to chase the bike (cf. Lohmann, this volume, Chapter 8).
8 As is true of Anutans, many Taumako have migrated to other parts of the Solomons, the most common destination being the area around Honiara on Guadalcanal. Their experience with dogs in Honiara also resembles that of Anutans.
9 "Chapter 7" and "Appendix 3", cited in this extract, refer to a chapter and appendix in Leach and Davidson 2008, and not the present volume.
10 Pyrek noted a pervasive conviction that "others" were not to be trusted. Tokelauans on their home islands have long lived in small communities with little privacy, and their day-to-day lives require trust in their neighbors. When

they leave their enclaves, they feel less safe. She found this particularly interesting because, from a Tokelauan perspective, Samoans—whom she found gentle and trustworthy—were the dangerous "others". Therefore, we were told, when Tokelauans move overseas, they are likely to take in dogs "for protection". Our interlocutors described Samoa as unsafe, with a "high crime" rate, including murder. Some Tokelauans expressed the view that in Samoa they face discrimination and are disproportionately victims of crime. One friend reported that he and his daughter had four dogs when they lived in Āpia. From a functional perspective, dogs may serve the purpose of mitigating fear of danger, especially on non-native islands, among "others".

References

Best, Simon. 1988. Tokelau archaeology: A preliminary report of an initial survey and excavations. *Bulletin of the Indo-Pacific Prehistory Association* 8: 104–118.

Clifford, James. 1988. *The predicament of culture: Twentieth century ethnography, literature, and art*. Cambridge, MA: Harvard University Press.

Clifford, James, and George Marcus, eds. 1986. *Writing culture: The poetics and politics of ethnography*. Berkeley: University of California Press.

Cramb, Justin. 2022. The Dogs of Remote Oceania: An archaeological and ethnohistorical view of domestic dog introduction and loss in the South Pacific. *Archaeology in Oceania* 57(1): 28–38.

Feinberg, Richard. 1979. Schneider's symbolic culture theory: An appraisal. *Current Anthropology* 20(3): 541–549.

Feinberg, Richard. 1981. *Anuta: Social structure of a Polynesian island*. Lā'ie HI and Copenhagen: Institute for Polynesian Studies and the National Museum of Denmark.

Feinberg, Richard. 1994. Contested worlds: The politics of culture and the politics of anthropology. *Anthropology and Humanism* 19(1): 20–35.

Feinberg, Richard. 2011. *Anuta: Polynesian lifeways for the 21st century*. Kent, OH: Kent State University Press. [Original, 2004]

Feinberg, Richard. 2019. A spiritual melting pot: Taumako, Solomon Islands. In *Haunted Pacific: Anthropologists investigate spectral apparitions across Oceania*. Roger Ivar Lohmann, ed. Pp. 90–117. Durham, NC: Carolina Academic Press.

Feinberg, Richard, Ute J. Dymon, Pu Paiaki, Pu Rangituteki, Pu Nukuriaki, and Matthew Rollins. 2003."Drawing the coral heads": Mental mapping and its physical representation in a Polynesian community. *The Cartographic Journal* 40(3): 243–253.

Feinberg, Richard, and L. Zimmer Tamakoshi, eds. 1995. Politics of culture in the Pacific Islands. Special issue, *Ethnology* 34(2–3).

Guillo, Dominique, and Nicolas Claidière. 2020. Do guide dogs have culture? The case of indirect social learning. *Humanities and Social Sciences Communications* 7(31): 1–9.

Huntsman, Judith. 2004. Tokelau exodus. *Journal of Pacific History* 39(1): 23–41.

Kroeber, Alfred L. 1952. *The nature of culture*. Chicago: University of Chicago Press.

Kroeber, Alfred L., and Clyde Kluckhohn. 1963. *Culture: A critical review of the concepts and definitions*. New York: Vintage Books.

Kroeber, Alfred L., and Talcott Parsons. 1958. The concepts of culture and of social system. *American Sociological Review* 23: 258–283.

Larson, Eric. 1966. *Nukufero: A Tikopian colony in the Russell Islands.* Eugene, OR: University of Oregon.

Larson, Eric. 1968. Tikopian labour migration to the Russell Islands. *Journal of the Polynesian Society* 77(2): 163–176.

Leach, Foss, and Janet Davidson. 2008. *The archaeology of Taumako: A Polynesian outlier in the eastern Solomon Islands.* Dunedin North: New Zealand Journal of Archaeology, special publication.

Parsons, Talcott, and Edward A. Shils, eds. 1962. *Toward a general theory of action: Theoretical foundations for the social sciences.* New York: Harper Torchbooks.

Schneider, David M. 1968. *American kinship: A cultural account.* Englewood Cliffs, NJ: Prentice-Hall.

Smith, Ian W. G. 2008. Appendix A3: Mammalian fauna from Taumako. In *The archaeology of Taumako: A Polynesian outlier in the eastern Solomon Islands.* Foss Leach and Janet Davidson. Pp. 401–405. Dunedin North: New Zealand Journal of Archaeology, special publication.

Titcomb, Margaret, and Mary Kawena Pukui 1969. *Dog and man in the ancient Pacific, with special attention to Hawaii.* Honolulu: Bernice P. Bishop Museum.

Tylor, Edward B. 1871. *Primitive culture: Researches into the development of mythology, philosophy, religion, art, and custom.* London: John Murray.

Whiten, Andrew. 2021a. The burgeoning reach of animal culture. *Science* 372 (6537).

Whiten, Andrew. 2021b. *The psychological reach of culture in animals' lives. Sage Journals* 30(3): 211–217.

Yen, Douglas E. 1973. Agriculture in Anutan subsistence. In *Anuta: A Polynesian outlier in the Solomon Islands.* Douglas E. Yen and Janet Gordon, eds. Pacific Anthropological Records, Number 21. Pp. 112–149. Honolulu: Bernice P. Bishop Museum Press.

7 Dogs without Their Humans

Occurrences of "Wild Dogs" in Oceania Prior to European Colonization

Loukas G. Koungoulos

Introduction

The dingo is the famous "wild dog" of Australia. Although dingoes were introduced to Australia by human maritime transport, and historically some were taken from the wild as pups and raised by Aboriginal people as companions, the vast majority predominantly live independent of humans and their resources. The bones of dingoes are quite common in Australian fossil deposits comprised of naturally accumulated animal remains (Gollan 1982) and included amongst these are some of the very earliest directly dated examples of the species, such as the Madura Cave and Koonalda Cave specimens which are over 3000 years old (Balme et al. 2018; Walshe 2017). It is generally accepted that dingoes have been living in the wild essentially as long as they have been in Australia, whatever their deeper history of domestication may be (Brumm 2021).

The dingo has a close cousin, known as the New Guinea Singing Dog in Papua New Guinea and the New Guinea Highland Wild Dog in West Papua (both henceforth referred to as Papuan highland wild dog or PHWD) (Cairns and Wilton 2016). This very rare animal constitutes the only other known verified example of truly "wild dogs"; defined as dogs who are not reliant on any form of human-derived assistance (direct or indirect) to maintain stable long-term populations, particularly with reference to meeting the demands of adequate nutrition and replacement-level reproduction (Koler-Matznick et al. 2003, 2007). Feral dog populations are reasonably common in many places but are differentiated by their continued reliance on anthropogenic resources for food and recruitment on domestic-dwelling dog populations to offset mortality. Both are distinguished from free-ranging domestic dogs who are fundamentally residents of human settlements and predominantly reliant on them for food and shelter, even if they engage in hunting and are not explicitly "owned" by a specific person or family. Due to their relatively independent lifestyles and lack of human selection over their reproduction, all three types may converge on the generic "pariah dog" phenotype which superficially resembles the dingo (Coppinger and Coppinger 2016).

DOI: 10.4324/9781003658696-7

Dogs were also introduced to most of the rest of Oceania during the late Holocene (Greig et al. 2018; Savolainen et al. 2004; Zhang et al. 2020). Outside of Australia and mainland New Guinea, the pre-Contact dogs of Oceania were often characterized by early European observers and commentators as docile, unintelligent and physically stunted varieties who lived entirely within domestic environments, kept under relatively restricted conditions by people who raised them on processed plant foods, essentially as livestock (e.g. Colenso 1878). Examples of dogs by this description are recorded from New Ireland (Studer 1901:35), the Tuamotus (Lucett 1851:247), Tahiti (Cook 1769, cited in Hawkesworth 1773b:10; Ellis 1831a:73), Hawai'i (Ellis 1831c:347), and New Zealand (Monneron 1769, cited in McNab 1914:287).

More detailed descriptions and archaeological evidence have supplemented this early image, showing that despite apparently deriving from a single founder population (Greig et al. 2018) the Oceanic dogs varied somewhat in size, shape, and pelage; they were sometimes involved in hunting (Anderson 1981), ate varied diets including terrestrial and marine meats, and otherwise often had important social roles or spiritual significance (Luomala 1960; Titcomb 1969).

Nevertheless the general picture of insular Oceanic dogs marks quite the contrast to the complex relationships of dingoes and PHWD to the indigenous peoples that tamed them from wild-born states, to live for a period of time as companions and hunting aids, and generally farewelled them back into the bush upon sexual maturity (Koler-Matznick et al. 2003; Koungoulos and Fillios 2020). The impression of many Oceanic dogs given by early European accounts is even in some ways different to the complex relationships between dogs and both coastal/pericoastal lowland New Guineans (Dobrin, this volume, Chapter 2) and Highlanders (Lohmann, this volume, Chapter 1) where the dogs' free-ranging natures and numerous economic, social, and spiritually-significant roles were perhaps more obvious.

The possibility of dogs having established wild or even feral populations in other parts of Oceania has not been seriously considered, perhaps for some of the reasons outlined above. This chapter aims to do so, in the hope of expanding our understanding of the diverse history of human–dog relationships in Oceania (Lohmann, this volume, Chapter 1). For brevity, I have limited discussion of potential wild dogs in other parts of Oceania to regions in which evidence was most forthcoming. I have attempted, where possible, to distinguish between the "feral" and "wild" categories as defined above. Dogs were not present in all Oceanic archipelagoes at the time of European contact, having either never been established there or having become locally extinct earlier (Cramb 2022). In these places any free-ranging or feral populations observed after the fact derive from those European dogs introduced post-Contact (Carucci, this volume, Chapter 9; and Prokosch, this volume, Chapter 3).

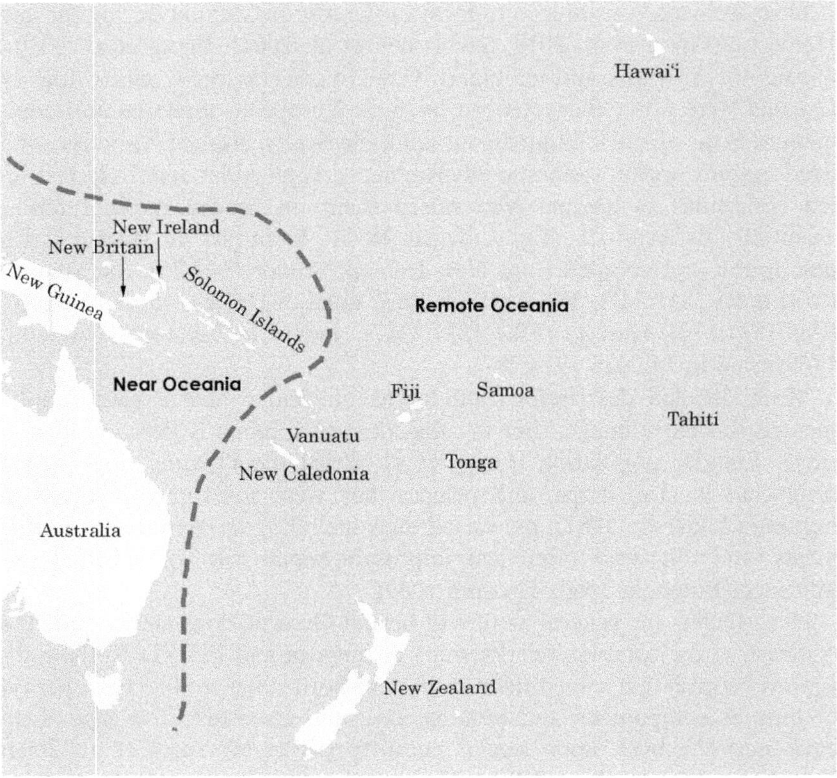

Map 7.1 Map of Near and Remote Oceania with key locations mentioned in the text. Map: Loukas Koungoulos.

Near Oceania

Dogs were first introduced to the part of Melanesia that falls within Near Oceania (an area defined by mainland New Guinea, its near-offshore island groups, and the Admiralty, Bismarck, and Solomon archipelagoes) by the Lapita people, a maritime culture originating in the Bismarck group, which settled across many of the south-western Pacific islands from about 3600 calBP. If strictly following radiocarbon chronology of well-dated specimens, Lapita-era dogs appear essentially simultaneously to the earliest directly-dated dingo remains (Balme et al. 2018). They are, however, extremely rare, invariably represented by small numbers of poorly-preserved elements, and could not have constituted a major or regular food source. Nevertheless, they appear in the Early Lapita context (3300–3000 calBP) at Kamgot in the Anir Islands (Summerhayes et al. 2019); in association with dates of 3220–3070 calBP at Brooker Island in the Massim group (Shaw et al. 2022); and in the earliest Lapita phase between 3234–3089 calBP and 3155–3020 calBP at Talepakemalai in the Mussau group (Kirch 2022:173).

Bismarck Archipelago

Some of the more convincing evidence for wild dogs outside of Australia-New Guinea comes from the Bismarcks. Descriptions of New Ireland's village dogs were given by Lesson, based on observations made in 1822 (Lesson 1827:123, translation from French cited in Dwyer et al. 2021):

> The dogs, named Poull, are small; their muzzle is pointed, and their ears are erect. They seemed to us in everything similar to those of New Holland. Courageous and very predatory, they live off everything they encounter, including fish and crabs, which they will fish on the reefs.

This dog was identified as identical to one Lesson later observed on an island off the north coast of New Guinea, probably one of the Biak Islands (Dwyer et al. 2021).

Another description of New Ireland village dogs in 1875 was given by Swiss ornithologist T. R. Studer (Studer 1901:34, translation from German cited in Gollan 1982):

> The dog is the size and has the habits of a middle-sized spitzer, only that the legs are longer, the hair shorter, smooth all over, even the tail, which is not annulate, and is normally bent a little upwards. The head is relatively wide, the snout sharply profiled, fairly pointed. The ears are wide apart, pointed and are upright. The colour is generally dirty white, or else spotted with dark, mainly black spots, or areas, often this black colour spreads over a large part of the body, so that only a white face "blaze", a chest spot, or a white throat and feet remain.

Although Lesson and Studer exclusively cover domestic contexts, small "wild" dogs were earlier observed in coastal forest of southwestern New Ireland by Carteret in 1767 (Carteret in Wallis 2016:1870), a sighting repeated, apparently in the same "uninhabited" general neighborhood, by Morell in 1830 (Morell 1832:455). That these individuals were simply free-ranging dogs from villages located some distance away, and unknown to these casual observers, remains quite possible.

It is on nearby New Britain that detailed description of wild dogs is more forthcoming. Ironically, descriptions of domestic dogs on this island are more scarce. However, they appear to be or have been broadly identifiable with New Ireland dogs, based on descriptions and photographs taken by archaeologist Ron Lampert on the Gazelle Peninsula in northeastern West New Britain (Gollan 1982:209–210) and various recent examples collated by Dwyer and Minnegal (2022a, 2022b).

During the 1960s at Umbi, in southern inland West New Britain, anthropologist Jane Goodale reported that wild dogs were hunted for food, and their pups taken for rearing as in Australia and the PNG

Highlands. Intriguingly, male village-born dogs were castrated by people, so free-roaming females were mostly impregnated by wild males. Given this intergenerational continuity, Goodale's (1995:70) description covers village and wild Umbi dogs alike:

> The dogs are small and terrierlike, with pointed ears, bushy tails, and short coats of varied colors. Informants say they have always been in the bush. There are a great many wild dogs in the mountains. One of the strongest memories I have of Umbi is the wolflike howling of the village dogs taken up by their wild cousins, the sound echoing back and forth, and then the sudden silence.

The howling of wild dogs was heard north of Umbi in 1932, near Mt Uali in the Whiteman Range by the AMNH Gilliard Expedition, naming this locality Wild Dog Ridge. Native informants expressed fear and claimed that the dogs belonged to "wild men" living in the mountains, though expeditionary members believed they actually escaped into the wilderness after the slaying of their owners in tribal conflict years prior (Gilliard and Lecroy 1967:179).

The wild dogs of New Britain do not appear to have been recorded in recent decades. Howling in the further-east Nakanai Mountains led members of a 2016 expedition, however, to give the newly-discovered Wild Dog Cave its name (Gabriel et al. 2018:8). More interestingly, a biodiversity survey in the same year in the Baining Mountains noted that mine construction had destroyed a cave in which local wild dogs had lived and bred; footprints and droppings indicated their ongoing residence in the area (Anjiga 2017).

Solomon Islands

Several notable observations were made in the main chain of the western Solomon Islands, after the annexation of this archipelago by the British Empire facilitated the first permanent European settlement and their explorations of the remote mountainous interiors of many islands. Dogs were seemingly introduced to this region sometime between 2000–1400 calBP (Wickler 2001) but potentially much earlier in parts of the Santa Cruz group, where they were present at Spanish contact in the late 16th century (Leach and Davidson 2008). We first hear from surgeon H. B. Guppy (1887:159–160) about Alu [Shortland Island]:

> Wild dogs are numerous in the bush in the interior of Alu. They never attack the natives or the pigs and, as they always slink away when alarmed, they are not often seen. They subsist on the opossums (Cuscus), waiting to catch them at the foot of the trunks of the trees as they descend to the ground at nightfall. When I was away on an excursion with Gorai the Alu chief, the native dogs that were with us

> ran down a wild dog and worried it to death … The unfortunate dog was apparently of the native breed.

On nearby Bougainville, anthropologist Richard Parkinson mentioned the hunting of "wild dogs, or dogs that have run wild" (Parkinson 2010:223) with spears and arrows alongside other game (in the 1880s–90s). The Halia language of northern Bougainville and Buka has separate words for domestic and wild dogs (Allen and Allen 1982).

Wild dogs were also found on Mala. Here, missionary W.G. Ivens (1927:403) described the local dog that was common during his early visits in the 1890s but had disappeared through admixture by the 1920s:

> The native dog of Mala and Ulawa was short haired, long in body, and with short and almost bandy legs. The ears stood erect and came to a sharp point … The tail curled over like a pug's tail, and clave to the body. Their bark was short, sharp, and staccato, very thin and high pitched. The howl was ear-piercing, and the whole of a village pack would join in.

The depredations of wild dogs caused terror amongst the local women and prevented keeping of chickens. Ivens (1927:54, 1930:246–247) believed that they were the result of village dog abandonment in the mountains after implementation of a dog tax, suggesting they were not true wild dogs. This event seems unlikely to be the sole factor in the existence of wild dogs as it was quite recent (1920; Hilliard 1966:196). These dogs were also apparently self-sufficient through hunting giant rats, cuscus, lizards, and land crabs (Ivens 1930:247).

Naturalist C. M. Woodford (1890:54) also described the native Solomons dog—apparently from Guadalcanal:

> The Solomon Islands dog … is assuredly a descendant of the dingo. I have seen individuals that in colour, shape of ears, general expression, and other characteristics, were hardly, except in size, to be distinguished from that animal.

Later climbing Mt. Popomanaseu, Woodford examined atop the adjacent peak of Mt. Makarakomburu several deep holes which his native guides identified as the burrows of resident wild dogs. They alleged a sixty-strong pack of these dogs had previously attacked and killed people before being dispersed by an organized hunting party (Woodford 1899:24). The dogs probably subsided on the giant rat *Uromys imperator*, which was noted as plentiful in the immediate vicinity, as is (today) the northern common cuscus *Phalanger orientalis*.

During WWII, soldiers on Guadalcanal became acquainted with the presence of wild dogs in the mountains. American soldiers credited their howls serving as advance warnings of air-raids (Burnett 1944:458), whilst

Japanese POWs reported having to eat wild dogs raw to avoid detection (Anon. 1943a). Wild dogs were encountered atop Mt. Popomanaseu's summit in the 1950s by geological surveyors such as J. C. Grover, who described them as "great" and "tawny" (Grover 1959:39). They were still seen until at least the 1970s (Hackman 1980:9).

Interestingly, even some of the smaller islands of the Solomons not part of the main chain, like Taumako, seem to have hosted feral dogs in their hilly interiors, though very little is known about them (Feinberg and Pyrek, this volume, Chapter 6).

Remote Oceania

Although demonstrably present in the Near Oceanian cradle of Lapita (Greig, this volume, Chapter 5), dogs were either not introduced or successfully established in Remote Oceania—the parts of the island Pacific beyond Near Oceania and largely corresponding to the cultural regions of Polynesia and Micronesia—by that culture. One exception is perhaps Fiji (see Cramb 2022), but here it is also unclear whether they were introduced immediately afterwards c.2800–2500 calBP in association with Polynesian Plainware pottery (Anderson 2009), or with a much later introduction c.1500 calBP (Davidson 2012). Proponents of the latter theory believe the dogs of Remote Oceania to have been associated with an intrusive cultural element from southeastern Micronesia, with the dogs originally having been obtained from a non-Lapita source in eastern Island Southeast Asia (Addison and Matisoo-Smith 2010). This scenario is supported by ancient DNA analyses of Polynesian dogs, which indicates separate sources/introductions for immediate post-Lapita dogs and later Polynesian dogs, in addition to the apparent pre-Lapita introduction of dingo/PHWD (Greig et al. 2018; Savolainen et al. 2004).

Samoa

Although the first two groups of European visitors to Samoa, led by Roggeveen (in 1722) and Bougainville (in 1768) do not mention native dogs, La Perouse's crew purchased two in 1787 in the Manu'a group (La Perouse 1798:219–236), and Kotzebue was traded one off Savai'i in 1824 (Kotzebue 1830:287). The next European visitor, missionary John Williams, was the first to actually settle in Samoa, on Savai'i, from 1830–1834. He (Williams 1837:498) provides the first description of a Samoan "wild dog":

> Another peculiarity in the natural history of the group is, that a wild dog is found in the mountains. I regretted exceedingly that I could not obtain one. From the description I received, it appears to be a small animal, of a dark, dirty grey, or lead colour, with little or no hair, and large, erect ears.

Williams also claimed that those slain in tribal conflicts were left "to be devoured by the wild dogs and cats of the mountains" (Williams 1984:245). Another missionary, J. B. Stair, on the neighbouring island of Upolu from 1838–1845, related that the Samoan dog had "run wild in the forests" and would occasionally descend in numbers into the villages looking for food. He described it as a "very shy" creature which he only glimpsed once amongst vegetation, at a distance. Otherwise, by this time, Stair relates that Samoan village dogs had become thoroughly mixed with European imports (Stair 1897:186) leading to the present-day Samoan dog population (Van der Ryn, this volume, Chapter 10).

Dogs in the Samoan mountains were mentioned again in the 1870s by A. B. Steinberger (1874:9), who noted the dog was unlike wolf or fox in appearance or behavior and "evidently" derived from domestic dogs "brought by shipping". Another 1870s visitor, Gordon Cumming (1882:69), gave a description of the animal almost identical to Williams' but adds that it had "short crooked legs" and a "long back", and that it subsisted on naturalized Pacific rat (*Rattus exulans*) and otherwise on breadfruit and yams. She claims that this information was determined when Samoa was "first discovered", but it is unknown where the aspects of her description differing from that of Williams were obtained. This is unfortunate, for those extra phenotypic traits give a strong impression of the Polynesian dog.

Society Islands

Dogs were observed in substantial numbers on Tahiti by several of the earliest European visitors beginning in 1767 (Commerson 1768:299; Wallis 1767, cited in Hawkesworth 1773a:188) and are well-known archaeologically on this island and throughout the Society Islands (Titcomb 1969). The Tahitian dog is the classic example of the diminutive, seemingly unintelligent and inactive Polynesian dog that was raised on a predominantly plant-based diet and primarily for human consumption. Instances of "wild dogs" in the Society Islands are limited to Tahiti, but are lacking in detail. The earliest of these, and perhaps the only useful one, is that of missionary William Ellis, who was in Tahiti from 1816–1822 (Ellis 1831a:70):

> There are no beasts of prey, nor wild animals, with the exception of a few boars or hogs, and dogs, in the mountains, and these are not often troublesome.

Like Williams, he also claimed that those slain in battle were left to be eaten by the "hogs or wild dogs, and the crabs" (Ellis 1831a:308).

Other sources seem to simply reproduce Williams' description of the Samoan dog in passages concerning Tahiti. The memoirs of very early missionary George Vason (1840), who visited Tahiti in 1797, contain a familiar-sounding description of a "wild dog" of a "dirty grey, or lead colour" with "large erect ears" and "little or no hair" (Vason 1840:27). This information, however, is not part of his "original narrative" and instead appears in a preface "introductory essay" written by the editor, the Rev. James Orange, who heavily references Williams. A similar explanation probably unfortunately applies to the statement of Gordon Cumming (1882:69) regarding a "wild dog" of the same physical description existing on Tahiti, since it is unclear where this information came from and her description of the animal is otherwise derivative of that of Williams for Samoa.

An additional issue is the relatively early proliferation of European dogs on Tahiti compared to many parts of eastern Polynesia. By the time of Cook's second visit to Tahiti in 1777, dogs "of two or three sorts" released earlier by a short-lived Spanish settlement had proliferated greatly and formed free-roaming if not feral packs who attacked livestock (Newell 2010:100–101). By the 1820s, the Tahitian dog had been wholly subsumed into a mixed-European population and was no longer identifiable as a distinct variety (Tyerman and Bennett 1832:284). Ellis' (1831a) reference to "wild dogs" in Tahiti was made shortly before this time, and he also notes that free-roaming dogs were contemporaneously a danger to livestock and people (Ellis 1831b:448–449).

Hawai'i

Dogs are well-attested archaeologically in the Hawaiian Islands from the time of first settlement, and a healthy population remained at first European contact by Cook in 1778 (Titcomb 1969). The so-called "poi dog" was essentially identical to the Tahitian dog from whom it probably derived. Several European and American ships visited the Hawaiian Islands each year from 1778 (Judd 1974), no doubt introducing new domestic breeds, but the poi dog appears to have survived alongside the tradition of dog-eating until the late 19th century (Luomala 1960; Titcomb 1969). The first reports of wild dogs, however, come with explorations of the mountainous interior of the "Big Island" (Hawai'i) after 1820.

The earliest of these is only an inferred reference to their presence as informed by others, not an actual sighting: J. F. Goodrich, the first European to ascend Mauna Kea in 1823, was informed that wild dogs had driven sheep (left decades earlier by Vancouver, which he observed) to flee to the summit (Goodrich in Ellis 1917:304). Two years later, botanist James Macrae made similar observations of dead sheep in the same general area, and was then held awake by the howling of wild dogs. The next day his party reached the hut of two European (feral) cattle-hunters who had recently seen a wild dog, giving the following interesting description (Macrae 1922:58):

> These men had seen no wild cattle since we left them, the only animals observed having been a wild dog and cat. The dog seemed to be the same kind as the domesticated native one of which they eat the flesh, and the cat appeared like the European breed.

Such specific identification with the indigenous poi dog is unique and not repeated in later reports, which lack phenotypic details. Poi dogs and their consumption were still very popular at this time, and the cattle-hunters, being individuals fully integrated into native Hawaiian society (their main dwellings being at a far lower altitude), must have been familiar with them. The Wilkes Exploratory Expedition later noted that in 1841 wild dogs were using lava tubes/caves to shelter on Mauna Kea's slopes (Wilkes 1849:106). By 1858, they were depredating the feral cattle and making attacks on a local turkey farm (Lyons 1945:49). Wild dogs continued to pose a major threat to cattle (Sheridan and Gartley 1911:404) and other livestock on Mauna Kea until after WWII (Maly and Maly 2006:236).

Hawaiian wild dogs are unique amongst Polynesian cases in that they were widely considered a threat to both large (bovid) livestock and to human safety in highland areas. The Wilkes expedition noted in 1840 the presence of bands of wild dogs in the uninhabited heights of Waimea Canyon in Kaua'i's interior 1840 (Wilkes 1849:70), who were reported to have killed native Hawaiians (Pickering 1848:94). Hostile packs of wild dogs were encountered by travel writers G. W. Bates on Mauna Haleakalā on Maui in 1845 (Bates 1854:323), and Mauna Loa (Hawai'i) in 1897 by J. R. Musick (Musick 1898:212–213). Dramatic warnings to tourists of wild dogs "almost as savage as the wolf" (Morris 1899:264–265) featured in travel books late as the 1980s (e.g. Bone 1983:30). Some recent cases probably derive from pig- and goat-hunting dogs that became lost or were abandoned (Morrison, this volume, Chapter 4).

New Zealand

Polynesian dogs (kurī) were brought to New Zealand by the first settlers, establishing a healthy population that was abundantly present in all regions at the time of first European contact in 1642. Admixture from European breeds led to its disappearance as a distinct form in most places by the 1850s (Luomala 1960; Titcomb 1969). Material for the consideration of the case of wild dogs in New Zealand is by far the most abundant within Oceania, and perhaps this is the only part of Remote Oceania where they appear specifically in oral traditions, but given limitations on scope only a summary of historical evidence is given here.

Wild dogs are first reported in 1814 in the Bay of Islands, North Island by traveler J. L. Nicholas. At this time only very brief (and unproductive) contacts with European ships had occurred and just three Europeans resided in the area (Nicholas 1817:126–127):

> On our return from the place where we cut down the spars, we met one of the native dogs, running about in a wild state. It was considerably larger than any of the dogs that we had seen domesticated among them, and bore a strong resemblance to the shepherd's dog so well known in England. The moment it came in sight of us, it set up a terrific howling; and never ceased the same baleful discord till we had left the place. There are numbers of dogs running wild in this manner through the different parts of the island; but I could not discover that they ever offered any injury to the inhabitants, who prize them very highly, as well for the sake of their flesh, which serves for a delicious article of food, as for their hide and bones, which they convert to a variety of purposes, in the way of ornamental devices.

Some kurī seemed to have a larger build which to some evoked the appearance of foxes (Crozet 1891:76; Cruise 1823:287), Indian pariah dogs (Dillon 1829:254), wolf-dogs (Lesson in Sharp 1971:82) and jackals (Dieffenbach 1843:45–46) and indeed the English "shepherd's cur" dog (Forster 2000:125).

It is possible some kurī led a more free-ranging and predatory existence compared to other Polynesian dogs (Greig and Rawlence 2021; von Haast 1871) though some doubt the breed was physically capable of doing so (Coutts and Jurisich 1973). In general, kurī were the most robust Polynesian dogs and had perhaps the most suitable base phenotype for such activity, particularly with regards to their crania, which exhibit far better developed biting- and holding-related jaw musculature than the Tahitian and Hawaiian varieties (Anderson 1981). By 1827 European strains had been added to the mix of free-ranging dogs in the Bay of Islands, obscuring any obvious traces of native descent (Earle 1832:147–148).

Reports of wild dogs became extremely common from the 1830s onward, as European settlement began in the South Island and intensified in the North Island. Sightings and encounters were documented publicly via newspaper correspondences, generating debate about the dogs' origins. Contemporary and later commenters were divided on whether they were kurī (Gillies 1877; Thomson 1921; White 1890, 1894), or newly-feralized European dogs (Anon. 1903a; Cheeseman 1893; Fairclough 1903; Hursthouse 1870), with occasional recognition some could have been predominantly European with a dash of kurī (Colenso 1878; Drummond 1913; Thomson 1859:21–22).

Some were clearly feral or stray European-origin dogs, originally working animals strayed from farmers and pest-controllers (Anon.,

1883; Anon. 1889; Anon. 1898; Anon. 1928; Hector 1872). However, a few of the more detailed sources depict phenotypes most consistent with kurī as described above (Anon. 1903b; Harris 1907; Hector 1876; C. G. 1890a, 1890b; Murison in Gilllies 1877:322–324). Several more might well describe kurī but are too vague in their detail to rule out European dogs (Anon. 1900; Anon. 1917; Gudgeon 1883; Wilson 1913). Interestingly, some apparently resembled dingoes (Torlesse 1851; Gillies 1877; Hook 1926).

Some believed that Māori dogs had adapted to the wilderness after abandonment (Dieffenbach 1843:86; Heaphy 1846 in Taylor 1959:242) or otherwise simply ranged freely/feralized out of Māori settlements (Torlesse 1858; von Haast 1861:135). They were notorious for killing sheep, but they seem to have primarily subsisted on native rats and flightless birds (Barrington 1864 in Taylor 1959:405; Brunner 1951; Dieffenbach 1843; Heaphy 1846 in Taylor 1959; Mackenzie 1894; Pearce 1897; White 1890). In addition, they were reproducing independent of people. One traveller found Māori collecting and rearing wild-born pups (Harper 1858); there are accounts of Europeans discovering wild-born litters (Anon. 1892; Anon. 1931), and then doing the same (Anon. 1911).

In the early 20th century reports of wild dogs noticeably decline in New Zealand newspapers. Although stories of feral dogs threatening sheep and occasionally people continue to appear in the media, they appear to be of entirely European heritage—often being traced back to lost pig-hunting dogs (Anon. 1943b).

Discussion

Historical accounts show that evidence for "wild" indigenous dogs is available for several larger islands in Oceania, with a few caveats. Firstly, in Remote Oceania the windows in which wild-living dogs specifically of such heritage are noted in apparent "wild" contexts are very brief, before the predominance of European strains forever compromises the picture (particularly for Tahiti). This is especially apparent when it concerns dogs that kill people and livestock, implying size, speed and aggression beyond what we understand of the native poi- and kurī-type dogs. It is possible that colonial-era dogs observed "living wild" and killing livestock but with phenotypic traits suggestive of native dogs were actually a mixture of the two, with the European proportion increasing over time (Greig and Rawlence 2021). It is also important to acknowledge that European colonization could have indirectly contributed to native dogs becoming wild-living, through the implementation of dog-taxes, forced relocations and other conflicts causing the abandonment of dogs outside domestic settings.

Secondly, in no case is it proven that Oceanic wild-living dogs were truly "wild" rather than "feral"—where the latter describes populations that are not self-sustaining in the long term, but continue to exploit anthropogenic resources and/or require ongoing recruitment from domestic populations to offset mortalities (Gompper 2013). The evidence *suggestive* of this is the consistent remoteness of most cases, being located in sparsely populated, high-altitude locales where replacement dogs might not have been readily available. The recorded instances of hunting carried out by these dogs are also suggestive that they were mostly self-reliant in terms of nutrition, although in some cases they appear to have been returning to villages for food (including "hunting" livestock). The existence of long-term feral populations of indigenous Oceanic origin, even if reliant on domestic recruitment, nevertheless constitutes a notable development in understanding human–dog history in the region.

Readers will notice that for a chapter concerned with pre-colonial occurrences little reference is made to ancient materials, which are particularly important in determining the likelihood that feral or wild dogs formed stable, long-term populations. Although there is not space here to provide a comprehensive assessment of all fossil-bearing sites in Oceania, a consultation of paleontological reports finds that in Australia the remains of wild dingoes are quite common in natural accumulations such as caves with pit-fall traps (Gollan 1982; Koungoulos 2022), but in the rest of Oceania dogs do not seem to occur in sites with strictly naturally-accumulated deposits. The reasons for this discrepancy are complex.

Firstly, some of the larger islands with geology amenable to the formation of caves or shelters with sediment traps such as pit-falls etc. happen to be locations for which there are no apparent records of wild or feral dogs, or even the presence of dog at all, prior to European contact. For instance, cave deposits with otherwise rich, non-volant vertebrate faunas on Grand Terre in New Caledonia (Anderson et al. 2010), on Viti Levu, Fiji (Worthy and Anderson 2009), and parts of western Micronesia like Guam for that matter (Pregill and Steadman 2009) all seem to lack dog. Conversely, some islands/archipelagoes with historic records suggestive of a pre-Contact wild dog presence—Savai'i and Upolu in Samoa and Tahiti are of most relevance here—do not contain many natural fossiliferous deposits, or the suitable sites for vertebrate preservation that do exist are not accessible to dogs by having entrances that are too small or too high. Here, and in many other islands, information about the native fauna of the locality is almost entirely derived from archaeological sites (Steadman 2006).

The example of New Guinea also shows that even where wild dogs are known to exist and have a wide geographic distribution, they may be paleontologically invisible in the absence of suitable accumulation sites. Naturally, here and elsewhere the extent of exploration, fieldwork and excavation in any given area plays a role in how visible such sites are. Nevertheless, in the Highlands canid remains in any kind of faunal (natural or cultural) deposit are incredibly rare, whereas in the lowlands they appear, albeit in relatively low numbers, in many Lapita and post-Lapita archaeological sites from as early as c.2800 calBP (Bulmer

2001; Manne et al. 2020). There is only a single example of a possible ancient *wild* dog specifically from New Guinea, at the Mapala Rock-shelter in subalpine West Papua (Hope 1976), which is located in the same area that today hosts a stable population of PHWDs (McIntyre et al. 2019).

Secondly, where sites physically accessible to dogs are present, these also tended to be used by people, and so where dog remains do occur in Oceanic caves with some natural deposits, it seems to be almost invariably in subsections of material attributable to cultural deposition. The best examples of this in Remote Oceania are Anatu Cave on ‘Eua, Tonga c. 500–2000 calBP (Steadman 1993) and Ana Pe’ape’a in American Samoa (Steadman and Pregill 2004), where considerable cultural material components (including other domestic animals from human meals) means that the canid remains cannot be reliably assigned to wild dogs.

Some isolated cases in Hawai‘i are interesting in this light; the volcanic geology of these islands features many lava tubes which can act as pitfall traps or as sinkholes and accumulate osseous material naturally. At least two such sites apparently contain dog: Māhā‘ulepū on Kaua‘i (Burney and Kikuchi 2006) and Barber’s Point on O‘ahu (Olson and James 1984:774). However, they also contain small amounts of cultural material (artifacts). Because the primary mode of accumulation seems to be that material is swept in from the surface by movement of water, it is probable that the deposits partly reflect redeposited archaeological material from nearby sites rather than strictly “natural” material. Indeed Māhā‘ulepū also contains remains of domestic ungulates probably associated with later European utilization of the surrounding landscape.

Conversely to the rest of Remote Oceania, dog remains are relatively commonly found in natural accumulations with no apparent cultural usage in New Zealand, which happens to contain the largest Oceanic landmasses outside of Australia-New Guinea. Greig and Rawlence (2021, table 1) have collated dozens of natural faunal accumulations from the North and South Islands, and even one on the Chathams, which contain dog. As in the Hawaiian sites, the problem of potential human association applies to these cases since they are typically located near to and are possibly associated with archaeological sites, and as such there is a possibility that these animals represent free-ranging dogs rather than a feral or wild population. Similarly, some of these deposits also contain European animals also introduced after Contact (such as sheep, cattle, horse, and brown or black rat, etc.) and thus the dogs could also potentially be representative of colonial-era feral European dogs. Nevertheless, the New Zealand specimens represent the greatest opportunity to verify the presence of a wild dog population in Remote Oceania.

The issue of fossil-capture and preservation aside, a possible general explanation for the discrepancy in historical or archaeological evidence for wild dogs in Near Oceania (Australia, New Guinea, larger islands in the

Bismarck Archipelago and some Solomon Islands) versus Remote Oceania may be found in varying degrees of suitability of different Oceanic environments to supporting self-sustaining wild dogs. Compared to Australia's diverse non-volant mammalian and reptilian fauna, the islands of Remote Oceania mostly lacked any substantial native terrestrial vertebrate prey sources (apart from introduced Pacific rat); ground-dwelling bird communities which may have otherwise offered suitable prey-sources suffered heavy extinction rates following initial human colonisations (Duncan et al. 2013). This leaves a selection of a few mostly small herpetofauna as the main vertebrate prey sources for feral or wild dog populations emerging after initial human colonization (Zug 2013).

It is notable in this light that the larger Near Oceanic islands, where the more convincing accounts of long-term native wild dogs come from, offer fairly sizeable terrestrial prey sources—where the northern common cuscus was supplemented by the small wallaby *Thylogale browni* in New Britain and New Ireland, and by lizards and terrestrial crabs in some of the larger Solomon Islands. Whether wild dogs in the Bismarcks and Solomons should be considered an extension of the range of the PHWD, or actually represent independently-derived establishments of other dogs, could be resolved through the DNA sequencing of these apparently still-extant populations.

The verification of legitimately native wild dogs occurring in Remote Oceania probably rests largely on New Zealand, although in principle any Oceanic examples of dog subfossil material found in sites formed by non-cultural accumulation processes offer potential for the same and deserve similar attention. A great deal could be learned through morphometric, genetic, and isotopic analyses of dog remains from New Zealand's natural sites, especially those with no attributable cultural affiliations, and also of historic wild dogs who upon capture were sent to various museums for preservation (e.g. Gillies 1877). Unlike much of the rest of Remote Oceania, in New Zealand dogs had a diverse offering of ground-dwelling birds to prey upon during Māori settlement, possibly contributing to many becoming extinct (Greig and Rawlence 2021).

Ultimately, whilst historical evidence cannot provide certainty about the presence of wild dogs analogous to dingoes and PHWDs having existed in other parts of Oceania, the consideration of such material here has certainly established the potential for promising further investigation into an under-appreciated matter.

References

Addison, David J., and Elizabeth Matisoo-Smith. 2010. Rethinking Polynesians origins: A West-Polynesia Triple-I Model. *Archaeology in Oceania* 45(1): 1–12.

Allen, Jerry, and Jan Allen. 1982. *Halia language, Halia to English, English to Halia, Dictionaries of Papua New Guinea*. Ukarumpa: Summer Institute of Linguistics.

Anderson, Atholl. 1981. Pre-European hunting dogs in the South Island, New Zealand. *New Zealand Journal of Archaeology* 3: 15–20.

Anderson, Atholl. 2009. The rat and the octopus: Initial human colonization and the prehistoric introduction of domestic animals to Remote Oceania. *Biological Invasions* 11(7): 1503–1519.

Anderson, Atholl, Christophe Sand, Fiona Petchey, and Trevor H. Worthy. 2010. Faunal extinction and human habitation in New Caledonia: Initial results and implications of new research at the Pindai Caves. *Journal of Pacific Archaeology* 1(1): 89–109.

Anjiga, Maxine. 2017. New species of orchids found in Papua New Guinea. *Medium*, 12/01/2017.

Anon. 1883. Wairarapa West County Council. Minutes. *Wairarapa Standard*, 14/05/1883, 2.

Anon. 1889. Local and general. *Wanganui Chronicle*, 15/04/1889, 2.

Anon. 1892. District & General. *Bush Advocate*, 09/07/1892, 2.

Anon. 1898. General News. *Southern Cross*, 09/04/1898, 9.

Anon. 1900. Untitled correspondence. *Evening Post*, 20/08/1900, 4.

Anon. 1903a. Our own animals. *Canis Maorium*. The pampered pet of the Maoris. A stupid and ugly "dumb dog". *Lyttelton Times*, 08/12/1903, 4.

Anon. 1903b. Untitled correspondence. *Poverty Bay Herald*, 07/03/1903, 2.

Anon. 1911. Local & general news. *Marlborough Express*, 21/04/1911, 4.

Anon. 1917. Southern pioneers. *Mataura Ensign*, 08/08/1917, 7.

Anon. 1928. In town and out. *Southland Times*, 30/06/1928, 8.

Anon. 1931. News of the day. *Auckland Star*, 18/04/1931, 8.

Anon. 1943a. Isle of Death. *Auckland Star*, 20/02/1943, 4.

Anon. 1943b. Wild dog menace. Packs in Taupo Country. *Gisborne Herald*, 27/04/1943, 4.

Balme, Jane, Sue O'Connor, and Stewart Fallon. 2018. New dates on dingo bones from Madura Cave provide oldest firm evidence for arrival of the species in Australia. *Scientific Reports* 8(1): 9933.

Barrington, A. J. 1864. Diary of a West Coast prospecting party. In *Early Travellers in New Zealand*. Nancy M. Taylor, ed. Oxford: Clarendon Press (1959).

Bates, George Washington. 1854. *Sandwich Island notes*. New York: Harper.

Bone, Robert W. 1983. *The maverick guide to Hawaii*. 7th ed. Gretna: Pelican Publishing Company.

Brumm, Adam. 2021. Dingoes and domestication. *Archaeology in Oceania* 56: 17–31.

Brunner, Thomas. 1951. *The great journey: An expedition to explore the interior of the Middle Island, New Zealand, 1846–8*. Christchurch: The Pegasus Press.

Bulmer, Susan. 2001. Lapita dogs and singing dogs and the history of the dog in New Guinea. In *The archaeology of Lapita dispersal in Oceania*. Geoffrey R. Clark, Atholl J. Anderson and Teresa Vunidilo, eds. Canberra: Pandanus Books.

Burnett, Wanda. 1944. What the fighting Yanks see. *National Geographic Magazine*, 451–476.

Burney, David A., and William K. Pila Kikuchi. 2006. A millennium of human activity at Makauwahi Cave, Māhā'ulepū, Kaua'i. *Human Ecology* 34(2): 219–247.

C.G. 1890a. Notes and Queries. *Otago Witness*, 20/011/1890, 23.

C.G. 1890b. Untitled correspondence. *Otago Witness*, 18/06/1890, 20.

Cairns, Kylie M., and Alan N. Wilton. 2016. New insights on the history of canids in Oceania based on mitochondrial and nuclear data. *Genetica* 144(5): 553–565.

Carteret, Philip. 2016. *Carteret's voyage round the world, 1766–1769*. Helen Wallis, ed. London and New York: Routledge.

Cheeseman, T. F. 1893. Notes on the New Zealand bats. *Transactions and proceedings of the Royal Society of New Zealand* 26: 218–222.

Colenso, W. 1878. Notes, chiefly historical, on the ancient dog of New Zealand. *Transactions and proceedings of the Royal Society of New Zealand* 10: 135–155.

Commerson, Philibert. 1768. Appendix 6. The Commerson documents. In *The Pacific Journal of Louis-Antoine de Bougainville, 1767–1768*. John Dunmore, ed. London: The Hakluyt Society.

Coppinger, Raymond, and Lorna Coppinger. 2016. *What is a dog?*Chicago: University of Chicago Press.

Coutts, Peter, and Mark Jurisich. 1973. Canine passengers in Maori canoes. *World Archaeology* 5(1): 72–85.

Cramb, Justin. 2022. The Dogs of Remote Oceania: An archaeological and ethnohistorical view of domestic dog introduction and loss in the South Pacific. *Archaeology in Oceania* 57(1): 28–38.

Crozet, Julien Marie. 1891. *Crozet's voyage to Tasmania, New Zealand, the Ladrone Islands, and the Philippines in the years 1771–1772. Translated by H. Ling Roth*. London: Truslove & Shirley.

Cruise, Richard Alexander. 1823. *Journal of a ten months' residence in New Zealand*. London: Longman, Hurst, Reese, Orme and Brown.

Davidson, Janet M. 1977. Western Polynesia and Fiji: Prehistoric contact, diffusion and differentiation in adjacent archipelagos. *World Archaeology* 9(1): 82–94.

Davidson, Janet M. 2012. Intrusion, integration and innovation on small and not-so-small islands with particular reference to Samoa. *Archaeology in Oceania* 47(1): 1–13.

Dieffenbach, Ernst. 1843. *Travels in New Zealand: With contributions to the geography, geology, botany, and natural history of that country*. Vol. 2. London: John Murray.

Dillon, Peter. 1829. *Narrative and succesful result of a voyage in the South Seas*. Vol. 1. London: Hurst, Chance & Co.

Drummond, James. 1913. The naturalist. *Otago Witness*, 12/03/1913, 68.

Duncan, Richard P., Alison G. Boyer, and Tim M. Blackburn. 2013. Magnitude and variation of prehistoric bird extinctions in the Pacific. *Proceedings of the National Academy of Sciences* 110(16): 6436–6441.

Dwyer, Peter D., and Monica Minnegal. 2022a. *Dogs in New Guinea: A collation. Part 1*. https://www.researchgate.net/publication/358165271_DOGS_IN_NEW_GUINEA_A_COLLATION_PART_I.

Dwyer, Peter D., and Monica Minnegal 2022b. *Dogs in New Guinea: A collation. Part 2*. https://www.researchgate.net/publication/359196384_DOGS_IN_NEW_GUINEA_A_COLLATION_PART_II.

Dwyer, Peter D., Harry Parnaby, and Monica Minnegal. 2021. A 19th century New Ireland dog, Canis familiaris novaehiberniae Lesson, 1827 and the status of Canis hallstromi Troughton, 1957. *Records of the Australian Museum* 73(4): 131–136.

Earle, Augustus. 1832. *A narrative of a nine months' residence in New Zealand in 1827*. London: Longman, Rees, Orme, Brown, Green & Longman.

Ellis, William. 1831a. *Polynesian Researches*. Vol. 1. London: Fisher, Son & Jackson.

Ellis, William. 1831b. *Polynesian Researches*. Vol. 2. London: Fisher, Son & Jackson.

Ellis, William. 1831c . *Polynesian Researches*. Vol. 4. London: Fisher, Son & Jackson.

Ellis, William. 1917. *A narrative of a tour through Hawaii, or Owhyhee.* Honolulu: Hawaiian Gazette Co.

Fairclough, P. W. 1903. To Manapouri, Te Anau, and Milford and back overland. *Otago Daily Times*, 09/02/1903, 2.

Forster, George, ed. 2000. *A voyage round the world.* Nicholas Thomas and Oliver Berghof, eds. Honolulu: University of Hawai'i Press.

Gabriel, Jennifer, Jim Specht, Mathew Leavesley, Matthew Kelly, Michael Wood, Simon Foale, Colin Filer, Susan McIntyre-Tamwoy, R. Michael Bourke, David Gill, and Jean-Paul Sounier. 2018. *The Nakanai Ranges of East New Britain, Papua New Guinea.* Cairns: James Cook University.

Gilliard, Ernest Thomas, and Margaret LeCroy. 1967. *Results of the 1958–1959 Gilliard New Britain Expedition 4. Annotated list of birds of the Whiteman Mountains, New Britain.* Vol. 135. New York: Bulletin of the American Museum of Natural History.

Gillies, Robert. 1877. Notes on some changes in the fauna of Otago. *Transactions and Proceedings of the Royal Society of New Zealand* 10: 306–324.

Gollan, Klim. 1982. Prehistoric dingo, PhD thesis, Australian National University.

Gompper, Matthew E. 2013. Introduction: Outlining the ecological influences of a subsidized, domesticated predator. In *Free-ranging dogs and wildlife conservation.* Matthew E. Gompper, ed. New York: Oxford University Press.

Goodale, Jane C. 1995. *To sing with pigs is human: The concept of person in Papua New Guinea.* Seattle: University of Washington Press.

Gordon Cumming, Constance Frederica. 1882. *A lady's cruise in a French man-o-war.* Vol. 1. Edinburgh: William Blackwood and Sons.

Greig, K., A. Gosling, C. J. Collins, J. Boocock, K. McDonald, D. J. Addison, M. S. Allen, B. David, M. Gibbs, C. F. W. Higham, F. Liu, I. J. McNiven, S. O'Connor, C. H. Tsang, R. Walter, and E. Matisoo-Smith. 2018. Complex history of dog (Canis familiaris) origins and translocations in the Pacific revealed by ancient mitogenomes. *Scientific Reports* 8(1): 9130.

Greig, Karen, and Nicolas J. Rawlence. 2021. The Contribution of Kurī (Polynesian Dog) to the Ecological Impacts of the Human Settlement of Aotearoa New Zealand. *Frontiers in Ecology and Evolution* 9.

Grover, John Charles. 1959. A geologist in the Solomon Islands. *United Nations Review* 5(7): 38–40.

Gudgeon, J. N. 1883. The ascent of Mount Tapaeanuka, the third highest peak in New Zealand. *New Zealand Herald*, 05/05/1883, 1.

Guppy, Henry Brougham. 1887. *The Solomon Islands and their natives.* London: Swan Sonnenschein, Lowrey & Co.

Hackman, Brian D. 1980. *The geology of Guadalcanal, Solomon Islands.* London: H.M. Stationery Office.

Harper, L. 1858. Exploration of the West Coast. *Nelson Examiner and New Zealand Chronicle*, 20/02/1858, 4.

Harris, J. L. 1907. The world's champion dog story. *Marlborough Express*, 15/08/1907, 7.

Hawkesworth, John. 1773a. *An account of the voyages undertaken by the order of his present Majesty. Vol. 1.* Dublin: J. Williams.

Hawkesworth, John. 1773b. *An account of the voyages undertaken by the order of his present Majesty. Vol. 2.* Dublin: J. Williams.

Heaphy, Charles. 1846. Notes of an expedition to Kawatiri and Araura, on the Western Coast of the Middle Island. In *Early Travellers in New Zealand*. Nancy M. Taylor, ed. Oxford: Clarendon Press. 1959.

Hector, James. 1872. Wellington Philosophical Society President's Address. *Wellington Independent*, 20/08/1872, 2.

Hector, James. 1876. On the remains of a dog found by Capt. Rowan near White Cliffs, Taranaki. *Transactions and proceedings of the Royal Society of New Zealand* 9: 243–248.

Hilliard, David Lockhart. 1966. Protestant missions in the Solomon Islands 1849–1942, PhD Thesis, Department of Pacific History, Australian National University, Canberra.

Hook, H. 1926. The Albertlanders. *Auckland Star*, 10/04/1926, 27.

Hope, Jeanette H. 1976. Fauna. In *The equatorial glaciers of New Guinea. Results of the 1971-1973 Australian Universities' Expeditions to Irian Jaya: survey, glaciology, biology and palaeoenvironments.* Geoffrey S. Hope, James A. Peterson, Uwe Radok, and Ian Allison, eds. Rotterdam: A. Balkema.

Hursthouse, Charles H. 1870. New Zealand's breadth and her wild dog. *Wellington Independent*, 08/09/1870, 2.

Ivens, Walter George. 1927. *Melanesians of the South-east Solomon Islands.* London: Kegan Paul, Trench, Trubner & Co. Ltd.

Ivens, Walter George. 1930. *The island builders of the Pacific: How and why the people of Mala construct their artificial islands, the antiquity & doubtful origin of the practice, with a description of the social organization, magic, & religion of the inhabitants.* London: Seeley, Service & Co.

Judd, Bernice. 1974. *Voyages to Hawai'i Before 1860: A Record, Based on Historical Narratives in the Libraries of the Hawaiian Mission Children's Society and The Hawaiian Historical Society, Extended to March 1861.* Honolulu: University of Hawai'i Press.

Kirch, Patrick Vinton. 2022. Vertebrate fauna from the Mussau Sites: Reptiles, birds, and mammals. In *Talepakemalai: Lapita and its transformations in the Mussau Islands of Near Oceania*. Patrick V. Kirch, ed. Los Angeles: UCLA Cotsen Institute of Archaeology Press.

Koler-Matznick, Janice, I. Lehr Brisbin, Mark Feinstein, and Susan Bulmer. 2003. An updated description of the New Guinea singing dog (*Canis hallstromi*, Troughton 1957). *Journal of Zoology* 261(2): 109–118.

Koler-Matznick, Janice, Bonnie C. Yates, Susan Bulmer, and I. Lehr Brisbin. 2007. The New Guinea singing dog: Its status and scientific importance. *Australian Mammalogy* 29(1): 47–56.

Kotzebue, Otto von. 1830. *A new voyage round the world in the years 1823, 1824, 1825 and 1826*. Vol. 1. London: H. Colburn and R. Bentley.

Koungoulos, Loukas George. 2022. The natural and cultural history of the dingo: A 3D geometric morphometric investigation, Ph.D Thesis, Department of Archaeology, University of Sydney, Sydney.

Koungoulos, Loukas George, and Melanie Fillios. 2020. Beyond ethnography and prehistory: the case of the Australian dingo. In *Dogs: Archaeology of the human-canine connection beyond domestication*. Brandi Bethke and Amanda Burtt, eds. Gainesville: University Press of Florida.

La Pérouse, Jean-François de Galaup, comte de. 1798. *A voyage round the world: Which was performed in the years 1785, 1786, 1787 and 1788, by M. De La*

Peyrouse: Abridged from the original French journal of M. de La Peyrouse, which was lately published by M. Milet-Mureau, in obedience to an order from the French Government. Edinburgh: J. Moir.

Leach, Foss, and Janet Davidson. 2008. *The archaeology of Taumako: A Polynesian outlier in the Eastern Solomon Islands, New Zealand Journal of Archaeology Special Publication*. Dunedin North: New Zealand Journal of Archaeology.

Lesson, René Primevère. 1827. Chapitre 2. Considérations générales sur quelques mammifère. In *Voyage autour du monde: Exécuté par ordre du roi, sur la corvette de Sa Majesté, la Coquille, pendant les années 1822, 1823, 1824, et 1825. Zoologie, Tome 1, Part 1*. René Primevère Lesson and Prosper Garnot, eds. Paris: Arthus Bertrand.

Lucett, Edward. 1851. *Rovings in the Pacific, from 1837 to 1849: With a glance at California*. London: Longman, Brown, Green and Longmans.

Luomala, Katharine. 1960. A history of the binomial classification of the Polynesian native dog. *Pacific Science* 14(3): 193–223.

Lyons, Lorenzo. 1945. *Makua Laiana: The story of Lorenzo Lyons, lovingly known to Hawaiians as Ka Makua Laiana, Haku Mele o ka Aina Mauna (Father Lyons, lyric poet of the Mountain Country)*. Honolulu: Honolulu Star-Bulletin.

Mackenzie, T. 1894. The West Coast fiords. Explorations between Manapouri and the West Coast Fiords. *Otago Daily Times*, 24/3/1894, 1.

Macrae, James. 1922. *With Lord Byron at the Sandwich Islands in 1825. Being extracts from the Ms Diary of James Macrae, Scottish Botanist*. Honolulu: W. F. Wilson.

Maly, Kepā, and Onaona Maly. 2006. *Hilo Paliku – Hilo of the Upright Cliffs: A study of cultural-historical resources of lands in the Laupahoehoe Forest Section, Ahupua'a of the Waipunalei-Mauluanui Region, North Hilo District, Island of Hawai'i (TMK Overview Sheet 3–7-01)*. Hilo: United States Department of Agriculture Forest Service—Institute of Pacific Islands Foresty.

Manne, Tiina, Bruno David, Fiona Petchey, Matthew Leavesley, Georgia Roberts, Katherine Szabó, Chris Urwin, Ian J. McNiven, and Thomas Richards. 2020. How long have dogs been in Melanesia? New evidence from Caution Bay, south coast of Papua New Guinea. *Journal of Archaeological Science: Reports* 30: 102255.

McIntyre, James K., Lisa L. Wolf, Benjamin N. Sacks, Johon Koibur, and I. Lehr Brisbin. 2019. A population of free-living highland wild dogs in Indonesian Papua. *Australian Mammalogy* 42(2): 160–166.

McNab, Robert. 1914. *Historical Records of New Zealand. Vol. 2*. Wellington: Government Printer.

Morell, Benjamin. 1832. *A narrative of four voyages to the South Sea, North and South Pacific Ocean, Chinese Sea, Ethiopic and Southern Atlantic Ocean, Indian and Antarctic Ocean, from 1822 to 1831 to which is prefixed a brief sketch of the author's early life*. New York: J. & J. Harper.

Morris, Charles. 1899. *Our island empire: A hand-book of Cuba, Porto Rico, Hawaii, and the Philippine Islands*. Philadelphia: J. B. Lippincott Company.

Musick, John R. 1898. *Hawaii, our new possessions: An account of travels and adventure, with sketches of the scenery, customs and manners, mythology and history of annexation to the United States*. New York: Funk & Wagnalls Company.

Newell, Jennifer. 2010. *Trading nature: Tahitians, Europeans and ecological exchange*. Honolulu: University of Hawai'i Press.

Nicholas, John Liddiard. 1817. *Narrative of a voyage to New Zealand, performed in the years 1814 and 1815, in company with the Rev. Samuel Marsden*. Vol. 2. London: Black & Son.

Olson, Storrs L., and Helen F. James. 1984. The role of Polynesians in the extinction of the avifauna of the Hawaiian Islands. In *Quaternary extinctions: A prehistoric revolution*. Paul S. Martin and Richard G. Klein, eds. Tucson: University of Arizona Press.

Parkinson, Richard. 2010. *Thirty years in the South Seas. Land and people, customs and traditions in the Bismarck Archipelago and on the German Solomon Islands*. Translated by J. Dennison and J. P. White. Sydney: Sydney University Press. Original edition, 1907.

Pearce, W. G. 1897. Acclimatization matters. *Feilding Star*, 19/06/1897, 2.

Pickering, Charles. 1848. *The races of man and their geographical distribution. Vol. IX, United States Exploring Expedition during the years 1838, 1839, 1840, 1841, 1842, under the command of Charles Wilkes, U.S.N.* Philadelphia: C. Sherman.

Pregill, Gregory K., and David W. Steadman. 2009. The prehistory and biogeography of terrestrial vertebrates on Guam, Mariana Islands. *Diversity and Distributions* 15 (6): 983–996.

Savolainen, Peter, Thomas Leitner, Alan N. Wilton, Elizabeth Matisoo-Smith, and Joakim Lundeberg. 2004. A detailed picture of the origin of the Australian dingo, obtained from the study of mitochondrial DNA. *Proceedings of the National Academy of Sciences of the United States of America* 101(33): 12387–12390.

Sharp, Andrew. 1971. *Duperrey's visit to New Zealand in 1824*. Wellington: Alexander Turnbull Library.

Shaw, Ben, Stuart Hawkins, Lorena Becerra-Valdivia, Chris S. M. Turney, Simon Coxe, Vincent Kewibu, Jemina Haro, Kenneth Miamba, Mathieu Leclerc, Matthew Spriggs, Karen Privat, Simon Haberle, Felicitas Hopf, Emily Hull, Alana Pengilley, Samantha Brown, Christopher E. Marjo, Geraldine Jacobsen, Lincoln Wesley, Robinson Nuabui, Starford Jubilee, Archy Losane Yapeth, Joe Norman, Paul Munt, Steven Lincoln, Isaiah Stanley, Eyasi Sanibalath, Tau Jack, Benard Isei, David Vilan, Robert Lincoln, Lincoln Inosi, Sima Lahaga, Wesley Lincoln, Tom Eliuda, Ernest Mark, Able Moimoi, Lemeki Isaia, Felix Jack, Heke Jack, George Sadiba, Solomon Ruben, Weda Gaunedi, John Sakiusa, Leon, Joseph Betuel, Kingsley, Ishmael, Edwin, Harry, G. Oscar, Joel, Jeremiah, Jimmy, Jerry, Roger, Joseph Nua, Lemeki, Nason, Thomas, Yadila, Brooker, and Papua New Guinea Panaeati Island communities. 2022. Frontier Lapita interaction with resident Papuan populations set the stage for initial peopling of the Pacific. *Nature Ecology & Evolution* 6(6): 802–812.

Sheridan, Solomon, and Alonzo Gartley. 1911. Mauna Kea, the highest of island peaks. *Mid Pacific Magazine*, December 1911, 403–411.

Stair, John Bettridge. 1897. *Old Samoa or flotsam and jetsam from the Pacific Ocean*. London: The Religious Tract Society.

Steadman, David W. 1993. Biogeography of Tongan birds before and after human impact. *Proceedings of the National Academy of Sciences* 90: 818–822.

Steadman, David W. 2006. *Extinction and biogeography of tropical pacific birds*. Chicago: University of Chicago Press.

Steadman, David W., and Gregory K. Pregill. 2004. A prehistoric, noncultural vertebrate assemblage from Tutuila, American Samoa. *Pacific Science* 58(4): 615–624.

Steinberger, Albert Barns. 1874. *Report on Samoa or Navigator's Islands made to the Secretary of State*. Washington, D.C.: Government Office Printer.

Studer, Theophil. 1901. Die prähistorischen Hunde in ihrer Beziehung zu den gegenwärtig lebenden Rassen. *Abhandlungen der schweizerischen paläontologischen Gesellschaft* 28: 1–137.

Summerhayes, Glenn, Katherine Szabó, Andrew Fairbairn, Mark Horrocks, Sheryl McPherson, and Alison Crowther. 2019. Early Lapita subsistence: The evidence from Kamgot, Anir Islands, New Ireland Province, Papua New Guinea. In *Debating Lapita: distribution, chronology, society and subsistence*. Stuart Bedford and Matthew Spriggs, eds. Canberra: ANU Press.

Taylor, Nancy M., ed. 1959. *Early Travellers in New Zealand*. Oxford: Clarendon Press.

Thomson, Arthur S. 1859. *The story of New Zealand*. Vol. 1. London: J. Murray.

Thomson, George M. 1921. *Wild life in New Zealand. Part I. – Mammalia*. Wellington: Marcus F. Marks, Government Printer.

Titcomb, Margaret. 1969. *Dog and man in the ancient Pacific with special attention to Hawaii*. Honolulu: Bernice P. Bishop Museum Special Publication.

Torlesse, Charles Obins. 1851. Mr Torlesse's report upon the Canterbury Block. *Lyttelton Times*, 05/07/1851, 7.

Torlesse, Charles Obins. 1858. *The Torlesse Papers. The Journals and Letters of Charles Obins Torless concerning the foundation of Canterbury Settlement in New Zealand, 1848–1851*. Christchurch: The Pegasus Press.

Tyerman, Daniel, and George Bennett. 1832. *Journal of Voyages and Travels by the Rev. Daniel Tyerman and George Bennett*. Vol. 1. London: Crocker and Brewster.

Vason, George. 1840. *Life of the late George Vason of Nottingham. One of the troop of missionaries first sent to the South Sea Islands by the London Missionary Society in the Ship DUFF, Captain Wilson, 1796. With a preliminary essay on the South Sea Islands, by the Revd. James Orange*. London: John Snow.

Von Haast, Julius. 1861. *Report of a topographical and geological exploration of the Western Districts of the Nelson Province, New Zealand*. Nelson: C. and J. Elliott.

Von Haast, Julius. 1871. *Moas and Moa Hunters*. Christchurch: The Times.

Wallis, Helen, ed. 2016. *Carteret's voyage round the world, 1766–1769*. London and New York: Routledge.

Walshe, K. 2017. Koonalda Cave, Nullarbor Plain, South Australia – issues in optical and radiometric dating of deep karst caves. *Geochronometria* 44(1): 366–373.

White, Taylor. 1890. On the wild dogs of New Zealand. *Transactions and Proceedings of the Royal Society of New Zealand* 22: 327–330.

White, Taylor. 1894. Te Kuri maori (the dog of New Zealand). A reply to the Rev. W. Colenso. *Transactions and Proceedings of the Royal Society of New Zealand* 26: 585–600.

Wickler, Stephen. 2001. *The prehistory of Buka: A stepping stone island in the Northern Solomons. Vol. 17, Terra Australis*. Canberra: ANU Press.

Wickler, Stephen. 2004. Modelling colonisation and migration in Micronesia from a zooarchaeological perspective. In *Colonisation, migration and marginal areas: A zooarchaeological approach*. Mariana Mondini, Sebastián Muñoz and Stephen Wickler, eds. Oxford: Oxbow Books.

Wilkes, Charles. 1849. *Narrative of the United States Exploring Expedition during the years 1838, 1839, 1840, 1841, 1842. Vol. IV, United States Exploring*

Expedition during the years 1838, 1839, 1840, 1841, 1842, under the command of Charles Wilkes, U.S.N. Philadelphia: Lea & Blanchard.

Williams, John. 1837. *A narrative of Missionary Enterprises in the South Sea Islands: With remarks upon the natural history of the islands, origin, languages, traditions and usages of the inhabitants.* London: John Snow.

Williams, John. 1984. *The Samoan journals of John Williams 1830 and 1832. Edited, with an introduction by Richard M. Moyle.* Canberra: ANU Press.

Wilson, A. 1913. The native dog (Kuri Maori). *The Journal of the Polynesian Society* 22(85): 42–43.

Woodford, Charles Morris. 1890. *A naturalist among the head-hunters. Being an account of three visits to the Solomon Islands in the years 1886, 1887, and 1888.* London: G. Philip and Son.

Woodford, Charles Morris. 1899. Colonial reports – annual. No. 275. British Solomon Islands. Report for 1898–9. In *Great Britain Parliament, House of Commons Accounts and papers.* Cambridge: Proquest LLC, 2006.

Worthy, Trevor H., and Atholl Anderson. 2009. Palaeofaunal sites and excavation. In *The Early Prehistory of Fiji.* Geoffrey Clark and Atholl Anderson, eds. Canberra: ANU Press.

Zhang, Ming, Guoping Sun, Lele Ren, Haibing Yuan, Guanghui Dong, Lizhao Zhang, Feng Liu, Peng Cao, Albert Min-Shan Ko, Melinda A. Yang, Songmei Hu, Guo-Dong Wang, and Qiaomei Fu. 2020. Ancient DNA Evidence from China Reveals the Expansion of Pacific Dogs. *Molecular Biology and Evolution* 37(5): 1462–1469.

Zug, George R. 2013. *Reptiles and amphibians of the Pacific Islands.* Oakland: University of California Press.

8 Cultural Differences in Dogs in Central New Guinea

Roger Ivar Lohmann

> dogs…share a significant number of characteristics, social practices, and habits with their humans.
>
> Yasmine Musharbash 2017:106

Introduction

As an ethnographer documenting the culture of the Asabano people in Papua New Guinea, I observed patterned differences in dogs' condition and interactions with people that differed from neighboring groups. Asabano dogs were skittish and thin compared to Telefolmin and Omai dogs. This seemed to indicate that contemporary, local cultures were distinctively expressed in these dogs' builds and actions. In describing these differences, I argue that humans are not the only species that express the distinctive cultures in which they participate. I also make the case that cultural diversity correlating with disparate multispecies community standards is expressed in dogs as well as humans.

During my stints of ethnographic fieldwork at Duranmin in central New Guinea, my attention was focused on the Asabano *people*. Nevertheless, I was struck by a few things about the local dogs.

I found the dogs living with Asabano people at Yakob and Daksil hamlets timid and standoffish toward humans. Interactions that I witnessed between them and their humans suggested why. Asabano people would, sometimes within minutes, express appreciation, indifference, and cruel intolerance to the dogs who lived, traveled, and slept in their company. If I were a local dog, I would have become cautious towards people, too.

During my main period of fieldwork from 1994 to 1995, I first acknowledged this impression when I visited other ethnic groups in nearby hamlets and found the dogs behaving differently than they did in my home hamlet of Yakob. In Telefolmin and Omai communities I immediately noticed that the dogs and people acted differently toward one another than did the Asabano people and dogs—and their dogs looked better-fed. These dogs looked and acted a little more like the dogs I was used to back home in North America.

DOI: 10.4324/9781003658696-8

Figure 8.1 Asabano dogs on the periphery. Downriver of Yakob, 2005. Photo: Roger Lohmann.

The thought grabbed me: were these differences an expression, in another species, of the distinctive cultures of the people these dogs lived with? Was this a case of cultural difference in dogs? If these dogs were expressing current local community cultures, in what sense was it *their* culture *as well as* that of their humans?

I found these thoughts striking because my anthropological education at the University of Wisconsin–Madison had emphasized culture—socially learned traditions—as a characteristic of humans. Distinct communities of people were expected to have different ways of thinking and acting; to have particular customs, to speak distinctively, to exhibit characteristic body treatments, and to make their own kinds and styles of artifacts. By comparison, since nonhuman animals lack the cultural and linguistic capacities of humans, they are expected to have much less, if any, regional or temporal variation and complexity of social learning and its outcomes. Moreover, the things that dogs learned in association with humans such as trick training, coordinated sled pulling, cooperative hunting, and elaborate command obedience were not usually called "culture" even though these were clearly socially learned and varied depending on dogs' communities. Cultural anthropologists have historically preferred to restrict the term to the elaborate symbolic systems that are more or less shared in human societies, although the term has been deployed in numerous ways since it entered 19th century anthropology (see Lohmann 2004 and in this volume Carucci, Chapter 9 and Feinberg and Pyrek, Chapter 6).

Anthropologists study people, and so they have also usually used the English word "communities" to refer only to humans. This restriction makes sense for certain purposes, but it can obscure our recognition of the other living things that participate collectively and interactively in their own ways. In its ecological sense, a community includes all species that are present in an ecosystem and interacting in competitive, predatory, parasitic, or mutualistic ways (Lambert 2012). Community ecology makes these multi-species groups the unit of study (Mittelbach 2011:2). Anthropology can profitably do the same, at very least for social participants like dogs. As Patricia McCormack (2018:138) observes, "Human societies do not consist of humans alone but involve an interspecies sociality in which dogs … may be present, sentient, exercise agency, and play roles in human lives". The ecological perspective also makes multispecies community relationships a point of reference, rather than restricting attention to a single species grouping. This makes interspecies cultural variation in dogs associated with different communities that include humans more recognizable and comprehensible.

Considering all this, I realized that the dogs of each culturally infused, multispecies community participate in its own distinctive culture. It should not have surprised me, for dogs are certainly influenced by how their people act toward them, which in turn varies in local cultures (Hecht 2017). More than this, as I argued in the introduction to this volume, dogs *manifest* parts of their human–canine community's culture that would not exist were it not for multispecies sociality. I call such expressions instances of "human–canine interspecies culture".

Some ethnic cultures depict other living things in an ecosystem—particularly hunted species—as fellow community members. These notions typically project imagined spiritual or infra-human qualities onto the other species (e.g. Brightman 1993; Brunois 2005; Willerslev 2007). For example, among the Rock Cree people studied by Robert Brightman in Manitoba, prey species were regarded as willing participants in a social exchange. They were supposed to repeatedly give their flesh and regenerate so long as people followed appropriate protocols of respect toward their bodies. I daresay that hunted animals, though they participate in culture-infused interactions with Cree hunters, have a different understanding that is nonetheless an outcome of these culturally specific interspecies engagements. Mythical notions about dogs as spiritual beings or metamorphosed humans also express ethnic cultural models of multispecies communities that affect how believers interact with them, and thereby what these dogs learn to think and do (e.g. de Vidas 2002 and in this volume, Dobrin (Chapter 2) and Van der Ryn (Chapter 10).

Eschewing supernatural and parochial beliefs, scientific community ecology helps us to more accurately see the multispecies communities of which various human ethnic groups are members. Anthropology helps us recognize that some of the cross-species interaction in our ecological communities is not merely predatory or mutualistic, but also, in fact, explicitly social and even cultural. In human–canine communities, both species join to manifest aspects of the community's distinctive culture.

The Asabano community seen in this light includes canids and primates sharing a montane rainforest habitat in central New Guinea.

Figure 8.2 Hunted pig in the foreground, hunting dog in the background. Yakob, 2007. Photo: Roger Lohmann.

These dogs and humans are but two of the (relatively recently arrived) species in a much greater and more intricate ecological community. The bodies, growth, and actions of other animals and plants, domesticated and wild, also express interspecies culture with the humans among them. However, as co-residents and co-carers, the paired populations of humans and dogs in Asabano, Omai, and Telefolmin cultural communities are distinctive in having established truly social ecological relationships with one another.

As "companion species", Donna Haraway (2003:11) writes, dogs "make life for humans what it is—and vice versa". Yet we two are not always companions, as the stable or recurrent doglessness of some of Oceania's islands shows (Feinberg and Pyrek, this volume, Chapter 6). Wild dog communities' separation from human communities has also occurred, evidently for a long time on larger landmasses (Koungoulas, this volume, Chapter 7). Where dogs and humans do co-reside, companionship does not seem to be universal or even necessarily common. This may reflect the discomfort and fellowship that are both potential reactions to our commonality as components of a greater unit: our species-transcendent dog-and-humankind. Dogs' distinctive relationships with humans "as almost-human or too-close-to-human" (Kondo 2021:8) informs not just cultural imaginings and beliefs about dogs as lovable, hateable, or edible (Herzog 2011:48–50), but also how humans interact with dogs such that distinctive cultural forms manifest in dogs' bodies and minds as well as in humans'.

Canines have not typically been considered a culture-bearing species. However, the picture looks a little different when canines are studied as members of culturally distinctive multispecies societies that include humans. Dogs can learn to understand words and signals, and carry out human commands (Tempelmann, Kaminski, and Tomasello 2014). They are capable of comprehending arbitrary symbols, and not just icons and indexes: so dogs, to a degree, share this symbolic ability commonly thought unique to humans (cf. Kohn 2007:5; and Musharbash 2017:98). The complex and precise responses of working sheepdogs to arbitrary commands demonstrates this; as Paul Keil (2015:508) notes: "the human handler forms with the canine partner an interspecies cognitive system". A cooperative integration is achieved. As Deborah Bird Rose (1992:176) observes of dogs living in the Western Australian Aboriginal community of Yarralin, "through living with humans, camp dogs become recipient[s] of human culture".

The enacted training of guide dogs has also been recognized as a form of culture transmitted by interspecific social learning (Guillo and Claidière 2020). Summarizing consensus in the literature on culture in nonhuman animals, Guillo and Cladière (2020:3) write that animal behaviors are recognized as cultural if four conditions are met:

> (i) the behaviour can be present in some populations and not in others; (ii) the acquisition of the behaviour is not fully explained by genetic differences between individuals; (iii) the behaviour must be socially transmitted … and (iv) an individual learning the behaviour increases the likelihood of other individuals learning the behaviour as well.

Applying these criteria to dogs, they argue that if dogs' human social partners are included as sources (and recipients) of socially learned information, then all four of these conditions are met. However, they exclude artifacts and living things influenced or created by culturally motivated behaviors as themselves cultural, because they do not themselves socially learn these characteristics. I find this exclusion unhelpful in understanding culture. Things and living bodies can encode and manifest cultural information and can be a link in the further communication of that information via learning to other minds. This is demonstrated, for example, by the successes of archaeology in reading cultural information from artifacts and written texts. Similarly, aspects of canine bodies and behaviors are effects of culturally motivated behaviors of people whose past breeding selection and interactions resulted in genetic and developmental consequences for dog lineages, groups, and individuals.

Both species participate in enculturation that is mutual and embodied, known as "structural coupling" (Foley 1997:12). Dogs and humans treat each other and members of their own species differently from one culture to the next, and both they and their humans bear the consequences of this different treatment in their bodies, behaviors, and knowledge. Nowhere is this clearer than in Lynn Morrison's (this volume, Chapter 4) account of the strikingly different human–canine interspecies cultures on different sides of the island of Hawai'i. Morrison documents health and emotional effects in both species that correlate with the different beliefs and relational standards found on the Kona side as opposed to the Hilo side.

Back in the 18th and 19th centuries in the same archipelago, Margaret Titcomb (1969:2) reports that the local Polynesian dog yapped rather than barked, but "evidently learned to bark very quickly after meeting European dogs". Hawaiian poi dogs were primarily raised by their humans as high-status food, and their diets and movements were restricted. These interspecies cultural conditions and not just their genes would certainly have contributed to their reputation among early foreign observers as sluggish and slow-witted.

Different cultural effects on dogs' physiques, habits, and abilities are widely visible in the world ethnographic record. While humans may direct, dogs are often recognized as participating in the training of one another to culturally appropriate tasks. Among the Northern Coast Salish of British Columbia in historical times, both humans and dogs who were experienced in the local ecosystem and hunting techniques trained new generations of hunting dogs; humans also intermittently interbred these dogs with wolves to enhance their hunting abilities (Anza-Burgess, Lepofsky, and Yang 2020). Romanian pastoralists' livestock guardian dogs are trained by exposure to sheep and experienced dogs. These working dogs' people do not play with them to reduce their distractibility, and crop tails and ears in the belief that it improves cautiousness and aggressiveness (Ivaşcu and Biro 2020:472–473). Sled dogs in Tierra del Fuego, Argentina, learn their jobs not just from humans, but "mainly because of the discipline demanded by other dogs (with barks, growls, looks, and bites) that they experience in practice" (Fanaro 2020:4). It follows that these cultures are neither purely human nor identical.

Garry Marvin (2005) provides an example of two radically different customary dog-and-human practices in different contemporary sectors of the same society. English Foxhounds, he writes, unlike English family dogs, are collectively owned. Rather than being cared for into old age, they are routinely shot by their loving keepers when no longer fit for work. Their training, living arrangements, and relationships with other dogs and people are tightly constrained according to distinct customs that appear strange even from the perspective of contemporary English pet-keeping conventions. If these aren't different human–canine interspecies cultures, what are they?

Figure 8.3 Asabano men and dogs gathered for a hunting expedition, Yakob 2007. Photo: Roger Lohmann.

What I Noticed about Dogs in Hamlets of the Om-Fu River Valley

One day in 1994, early in my most lengthy period of fieldwork, I recorded the activities around me as I sat in the Asabano communal house at Yakob village—a rectangular, one-room wood and leaf-thatch building on stilts with four hearths. A dog entered the house looking for food. He scratched at fleas as he sniffed about, wandering from one pile of raw sweet potatoes to another, finding nothing he wanted. Since I had often seen dogs struck when they approached looking for food, I was surprised that this dog moved unmolested by the people sitting here and there on the floor, preparing to cook their garden produce in the ashes for breakfast. The pattern that my observations had suggested so far was that when food was around, a dog or two would mill about the house wagging their tails cringingly, only to be swatted to a shrill yelp by any person they came near. These hungry, skinny dogs were remarkably persistent, and Asabano children soon learned from adults to shout "aiehh!" and swat the dog with a stick or even the dull of a knife, drawing an indignant yelp. No one was hitting this dog this morning, I learned, because they knew he would not take uncooked sweet potatoes.

The harsh slaps and resulting yelps pained me to witness. Though focused on the people, I also sympathized with the dogs. Immersed in my evening reading in the field, I was impressed to see similar treatment of dogs recorded among 19th century Lake Superior Ojibway (Kohl 1985 [1860]:37–38):

> The whole day through the poor brutes are heard yelping, and, altogether, they have a very seedy appearance. Their great object of life is to crawl into the huts and carry off something eatable; but they are continually driven out by the women and children, and recommended, by a smart blow, to satisfy themselves with the fish and beaver bones thrown out for them .... If you attempt to pat one of these dogs, and speak kindly to him, he does not understand what you mean; he tucks his tail between his legs and slinks off to a convenient distance, where he begins growling.

On the following page, Kohl explains that his Ojibway informants reckoned dogs unclean and treated them with cruelty, and yet also regarded them as holy and fit for sacrificial offerings because of their great usefulness. There is an ambivalence here in how humans treat dogs—and in how dogs treat humans—that is reminiscent of Asabano human–dog interactions and appears in other ethnographic contexts as well (e.g. Holland 2021). Although Asabano people did not sacrifice or eat dogs, despite what looked to me like unkindnesses, they admired the services of good hunting dogs (who also evidently acquired for themselves more to eat than others). Describing historic Polynesia, Katherine Luomala (1960:190) writes:

> The native attitude toward the dog was ambivalent. On the one hand the dog was a symbol of prestige, a desired accessory for human beings which improved their status. On the other hand the dog was a symbol of the social outcast, a despised and often comic hanger-on of human society.

Both Asabano humans and dogs looked underfed in the mid-1990s (although better ten years later), but clearly the dogs were contributing to the general food supply as hunting partners. Early on I saw a bandicoot being cooked. I was told it had been killed by a dog named Puppy while hunting with Keli, a middle-aged man. The dog had been named with this English word by his boy owner, Luke, who probably learned it at Duranmin Community School. When men went into the bush for a hunting expedition, they took one or more dogs with them (cf. Plata and Montiel 2020).

The food-shelter-and-protection bond is evidently important to both dogs and humans in their joint communities. However, the fact that dogs remain with particular humans even when poorly treated and provisioned suggests that social belonging itself, craved not just by humans but also by wild and domesticated canids, is a major source of the bond between dogs and their humans (see Payne, Bennett, and McGreevy 2015; Beetz et al. 2012:2; Wynne 2019).

Figure 8.4 An emaciated dog on a verandah in Yakob, 2007. Photo: Roger Lohmann

The free-living dogs that Koungoulos (this volume, Chapter 7) surveys group themselves into societies and socially learn skills distinctive to their backgrounds and opportunities, too. Some of these dog tribes live independently in high-altitude fastnesses, while others orbit around packs of wild humans. Dogs, like humans, choose to associate rather than going it alone; even in dysfunctional relationships (Canejo-Teixeira et al. 2020). Both species often put up with ambivalent treatment, domestic abuse, and worse from others in their societies to maintain membership in family or ethnic groupings. In the Trans-Fly region of southeastern New Guinea, pioneering ethnographer F. E. Williams (1936:421) describes such ambivalent interactions between the two species in their community:

> The Moorhead dogs suffer from mange and other diseases that reduce them to a wretched condition. Instead of being mercifully killed they are simply ignored, or, when with the desperate tenacity of starvation they intrude themselves wherever food is to be seen, they are struck and kicked out of the way with a kind of thoughtless cruelty. And yet at other times a man may show great fondness toward them. I have seen a youth … boisterously welcomed by his dogs; and he did not fail to return their show of affection.

On another 1994 morning in Yakob village, I was moved to record in my fieldnotes the "incredible inconsistency" with which dogs were treated. A puppy, peacefully sleeping, was kicked out of his sleeping place. He sleepily wandered in one direction, when a young woman grabbed him up so harshly that he yelped. She pulled the puppy to her lips and kissed him and then threw him in the direction of the doorway with equal ardor.

I found the juxtaposition of affectionate and harsh treatment of a puppy—a child dog—within seconds surprising and unsettling, but it suggested the kinds of cultural conditions that puppies are trained by when growing up as a dog in Asabano society. Recalling this incident reminded me of Margaret Mead's description of Balinese mothers training their babies to be impervious and emotionally controlled by juxtaposing kind enticements with taunts and disengagement (Bateson and Mead 1942:32–33). Culture and personality research ultimately established that child-rearing practices do not translate to something like typical personalities in a culture (Bock 1999:116). Nevertheless, there is no doubt that socialization (learning a group's social behavioral standards by participation) and enculturation (incorporating cultural information of all kinds by participation in a group) more broadly do channel society participants' models of and for reality, including the behavioral and physical expressions of those models. Members of a society learn common expectations from the people and things around them, and strategies for successfully navigating such conditions (Goodenough 1981). Furthermore, cultural conceptualizations of dogs as dangerous or cute, as family or tramp (as

Jordan Prokosch points out in this volume, Chapter 3), take the form of schemas that are expressed in actions. When dogs in Asabano society routinely encounter inconsistent treatment from humans, it seems reasonable to suppose that they learn to behave around humans in a hopefully fearful, skulking manner like what I observed.

Another of Mead's observations on Bali exemplifies how dogs enact their society's particular culture:

> at the New Year every sound is silenced. Even the dogs, which on every other day keep up a sharp, impersonal yapping at every passer-by, sense the need for silence and skulk away into the courtyards where each family, with fires out and offerings set out in the house temple, stays quietly by itself
>
> (Bateson and Mead 1942:2).

After living among the Runa of Ecuador for several months, ethnographer Francesca Mezzenzana (2023:197) was horrified when her three-year-old son absorbed one local convention and started hitting with sticks and stones dogs who appeared to be "lazy" rather than good hunters and providers. In central Australian Warlpiri society, Yasmine Musharbash (2017:97) found that adults encouraged toddlers to roughly abuse puppies. "This 'cruelling' of puppies by toddlers instils in both Yapa [Warlpiri people] and camp dogs the notion that Yapa, including the very young, are 'stronger' (negating the actuality of dogs' superior strength, especially of the jaw)". Possibly, some of the inconsistent treatment of Asabano dogs that I observed had similar causes and consequences. However, it appeared in the main to be "done relatively thoughtlessly, that is, with a relative lack of empathy" (Lohmann 2011:107).

Some Asabano village dogs were terribly thin, had skin infections, and were ignored or swatted away by the humans in their community. Others, particularly prized for their aid in finding and fighting with prey, were called with a whistle to join in hunting expeditions. I noted four dogs at an Asabano garden house called Kalex (an informal form of Kale *Melema* or settlement) in 1994. The dogs taken to this remote location, surrounded by forest, would likely have been relatively favored as hunting partners. These dogs were so noticeably different from those I had observed in the main hamlet of Yakob that I remarked in my fieldnotes that they were "better fed and better treated than the average village dog". Nevertheless, they were not spared the rod. After a disturbed night spent there with human and canine companions, I grumpily described the scene in my fieldnotes:

> Didn't enjoy sleeping with everyone last night. Interruptions of talking, putting more wood on the fires, babies screaming periodically. People

> sleep on a piece of bark near the hearths, with no covering other than their clothes or a piece of cloth. They often lie on their backs, with their legs draped over the bush rope guards around the hearth, about half a meter high. Talking begins intermittently at about 5:00 am [an hour before sunrise], with no concern to be quiet—full voices are used as always. [In contrast,] When dogs make the slightest sound, from whining lightly while scratching lice to howling [singing—the native dogs are New Guinea Singing Dogs], they are sharply reprimanded by human words that are much louder, or by a smart swipe with a stick, producing a much louder yelp.

At other times I saw Asabano people express love to their dogs, too. We social animals are perhaps all too prone to putting up with and dishing out unkindness. One evening in 1995 I began a pleasant night in the newly constructed men's house. Elders Bledalo and Salua, a young man named Waiye, and several little boys were present. Shortly after I arrived, Madfe—a Paiyamo refugee of long-ago tribal warfare—offered me some cooked sweet potato, and then gave some to a young, friendly dog named Spiky. The fact that I found remarkable this feeding of a young dog, and the outgoingness of that dog, indicates the unusualness of the sight for me. The relative friendliness that I noticed in this puppy, accompanied by his loving treatment and feeding by Madfe, foreshadowed experiences of dogs in nearby non-Asabano communities.

In 1995, I visited Kusup hamlet, inhabited primarily by Hewa-speaking Omai (Mondubanmin) people near the home territory of the Lembana people some days' walk downriver from Duranmin in the Om River valley. While there, I talked to an Omai man named Matthew about hunting with dogs. He told me he had seen tree kangaroos defending themselves from attacking dogs by hitting them with sticks grasped in their hands for the purpose. Upon hearing this I excitedly contemplated the possibility of marsupial tool use with visions of Jane Goodall's iconic discovery that chimps prepared sticks to fish for termites dancing in my head. Then I noticed a big difference in how the Omai dogs acted. I remarked in my fieldnotes, "Dogs here behave differently than at Duranmin. They come up to people without fear, touch them, and they are fatter. People are much kinder to them here". A similar observation soon after provoked the same reaction in me. I was visiting Brengam, a Telefolmin hamlet on the Fu River, a couple hours' walk upriver of Duranmin airstrip. I recorded in my fieldnotes, "Here they are friendlier to dogs, and the dogs are not afraid". I also recall noticing that the dogs living with Telefolmin people at Ilimdang hamlet, immediately across the Ilim Stream from the Duranmin airstrip, behaved more like dogs back home in Wisconsin—coming up to people without fear, and looking better fed.

Map 8.1 Duranmin Airstrip, Sandaun Province, Papua New Guinea, New Guinea Island.

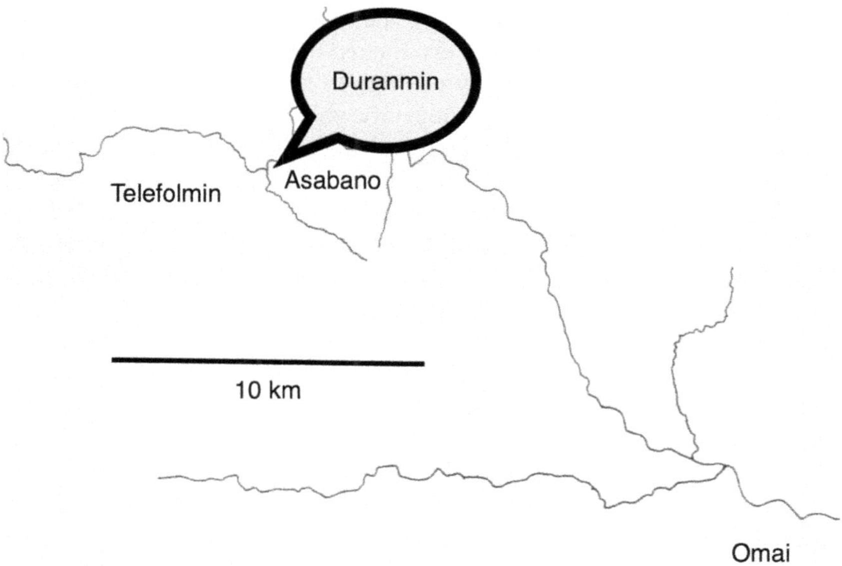

Map 8.2 Asabano, Telefolmin, and Omai groups relative to Duranmin Airstrip (scale approximate).

Lyle Steadman (1971:50) describes dogs (singular: *yao*) among Hewa speakers in 1968 interacting in a manner that is more in keeping with my observations of Asabano dogs than was current in 1995 among the Hewa-speaking Omai of Kusup:

> A young dog is trained simply by being taken on hunts with an experienced *wam* [wild animal hunting] dog. If the dog does not show any hunting potential, it is simply ignored and although tolerated in the house, it is beaten much more readily for trying to steal meat or cooked sweet potato than a wam dog. No dogs are fed, whether they are good hunters or not. Consequently, they are always looking for an opportunity to steal food. They also readily cringe, and, baring their teeth, cry like a young child whenever anyone picks up a stick. Their food consists mainly of whatever they catch in the forest or manage to gulp down before their owner wrenches it away from them. In addition, they eat discarded peelings of sweet potato which they cautiously retrieve from the fire-pit, as well as discarded animal bones. The only time a dog may receive some food from its masters is when it is still a puppy. During this time, it may be fed small bits of cooked sweet potato. Although of small amount, this may be the critical factor which attaches the dog to man.

Describing Telefolmin dogs, Jorgensen (1981:74) writes about their nearby heartland of Ifitaman:

> Most men keep dogs (*kayaam*), small and usually skinny beasts, which can—especially if well looked-after—help them in their search for game. The ability of dogs to sense or detect animals in the bush is much-admired—dogs are able to do this because when Afek ["Old Woman," the mythical ancestress of both Telefolmin and Asabano] scattered all the wild animals far from the village of Telefolip, the men covered their eyes (as they were supposed to) while her dog peeked. Though not considered nearly as valuable as pigs, nor as well cared for, a man often feels great affection for a good dog, who will often accompany him whenever he leaves the village.

In the Telefolmin myths Jorgensen (1980:516) collected, a dog accompanied the founding Old Woman in her journeys, and he notes a possibly "widespread association between primordial women and male dogs" citing research in other parts of Papua New Guinea. Jorgensen's Telefolmin informants also believed in a race of cannibal amazons reputed to marry also-greatly-feared wild dogs (*dunkiil*) who live in remote wilderness areas, suggesting an intriguing distinction in Telefol thought between dogs who are enculturated as part of human society as opposed to dogs (and other beings) who are outside of it and its controls (Jorgensen 1980:93). Some groups claimed descent from domestic or wild dogs, the latter of whom reputedly taught humans the art of fire-making—a case in which dogs are believed to have enculturated humans, and to be of interchangeable substance with them (1980:160, 162). In the local pre-Christian religious system, killing and eating dogs had been forbidden under normal circumstances, but this taboo, like others, was lawfully broken as part of secret male ritual (Jorgensen 1980:11, for more details see Jorgensen 1981:126–128).

The word for dog in Asaba is /alodu/. I collected all manner of common phrases including some that provide a window to the kinds of things Asabano people of the time were saying about (and to) dogs:

> /Alodu numa meole yeyamodu./ = Dog came in the house looking for food.
> /Alodu abadiene!/ = Expel the dog!
> /Alodu abe!/ = Dog, get out!

The following phrases, in contrast, address the positive associations Asabano people have with dogs as hunting aids:

> /Adi alono abedele oba./ = We two take the dogs and go. (Let's go hunting with the dogs.)
> /Adi alono abedele oma, wobusi suma odema./ = We two get dogs and go to the bush.
> /(Adi) Alono delem demono abeyoba./ = (We two) Get dogs and are going hunting animals.

Asabano dogs sometimes get into tussles with one another, and Asabano people do not approve.

> /Alono ohisano./ = The dogs fight.

In 1995, Belok told me that like pigs, "We also castrate most male dogs, and just leave a few to impregnate the others. If we left all the male dogs whole, they would constantly be fighting over females. Castrated dogs will still fight with [prey] animals". Asabano people highly praise this latter sort of fighting in dogs. Before their conversion to Christianity in 1977, they fed their hunting dogs scrapings from a femur supposed to have belonged to a sprite to make them more vicious.

Hunting dogs use their superior sense of smell to locate and lead their human hunting partners to prey. Dogs may kill smaller prey like marsupials and harry large prey like wild pigs and cassowaries until a human hunter in the party shoots the unfortunate quarry with a bow and arrow, or in some more recent cases, with a rifle. On hunting expeditions with their humans, dogs take on marsupials smaller than themselves, such as cuscuses, ones of similar size, like tree kangaroos, and prey that are much larger than themselves, including cassowaries and pigs who are capable of killing both dogs and humans. In one account, dogs who fought valiantly with a pig who nearly killed his human attacker were credited with saving the man's life.

While Asabano men are the primary hunters, women may engage in opportunistic hunting, sometimes occasioned by companion dogs. One morning in 1994, a young married woman named Tona came to my house and gave me a piece of pork. She and her sister Aune had been the far side of the Fu River with a dog, and their dog began fighting with a fairly large pig. Tona came up with her axe and killed the pig. They made a steam oven and cooked it on the sandy bank of the river. "I'm no man," she said; "I didn't use a bow and arrow, but I killed it with my axe."

Once or twice in the mid-1990s I heard people speak with approval of the reputed prowess of hunting dogs from other parts. One man for example, knowing my interest in traditional drums, offered to give me one if I would buy him a hunting dog when I visited the provincial capital, Vanimo—an offer I did not follow up.

Dogs frequently accompanied families when they journeyed away from the central hamlet for expeditions to remote gardens, to hunting shelters in the bush, or to harvest in stands of sago. One day in 1994, for example, I noted that upon visiting the family house of Sokiale and his wife Kalifeyo, I found it empty and was told that the family, including two or three children and three dogs had left that morning for a two-day walk down the Fu River to cut sago. The dogs, my informant offered, were of three generations.

People not only paid some attention to their dogs' family relationships; they also saw them as susceptible to the same supernatural dangers that beset people. In 1994 Yalowad, a father of teenagers, said that a mother and child dog of his were both killed by a cannibal witch who had turned into a large yellow poisonous snake while they were hunting together in the forest. "The dogs fought with the snake, and it bit them both and they died very quickly," Yalowad said. "I hit the snake with a stick and broke its head and left all three to rot." Leaving good meat to rot would be anathema for Asabano hunters, which indicates that Yalowad considered neither the snake nor the dogs edible. Although at least some snakes were eaten, the impression I had was that this snake's purported identity as a transmogrified cannibal witch had made it unappetizing. I did not hear of people ever eating dogs. This was in part because they were valued for hunting, trade, and breeding, and in part because they were regarded as unclean because of their habit of eating feces and, in former times, rotting human meat that fell from tree platform graves. Yalowad remarked that a puppy of his had also died recently. A child who was present said it was killed by a pig, but Yalowad quickly added that it was cannibal witchcraft—the puppy had just a small hole in its belly, but a lot of its insides were coming out while it was still alive.

Asabano people supposed that dogs could not only be victimized by cannibal witches, but could also, like other animals, themselves *be* human cannibal witches in shapeshifted form. In 1994, a young man named Jim Alosi remarked:

Figure 8.5 Tona chose to pose for her picture with a puppy and a baby in Yakob village, 1994. Photo: Roger Lohmann.

> If you go alone to the forest and you hear a puppy or dog barking on your path, it's following you. Then you'll be afraid, and all your hair will stand up and you'll know it's a cannibal witch following you. You will know cannibal witches are there by your fear.

The idea that dogs are somehow humanlike yet not humans in their intensive social interaction with their people is recognized in supernatural notions that they can be victims *or* perpetrators of cannibal witchcraft—in other words—they can be the bodily forms *and* the meat of the mythical cannibal witches that figure large in traditional Asabano understandings of the causes of death. To me, at least, the howling of New Guinea village dogs could sound eerily like people's voices. This may reflect their common heritage with free-living New Guinea Singing Dogs (Dwyer and Minnegal 2016, 2021). On one hot afternoon in 2005, I was sitting with a young man named Laurie when strange sounds that I could not identify emanated from Yakob village. I asked him what the sound was, and he said, "It's the dogs howling because everyone has left the village and they are worrying."

In 2007, I visited a settlement that was actually named for the dogs who inhabit it. Belok and I walked the 10–15 minutes from Yakob to an isolated house on top of a breezy hill overlooking the Fu River. He called it *Alono Mele adema*, "dogs' settlement" or Dogs' Corner (*kona* is Tok Pisin for a tiny hamlet of one or two houses). The house had two rooms and hearths. One room housed Belok's elder sister, Belok, his wife Omsimin, their kids, and Omsimin's parents Yalowad and Waloko. The other room was for boys to sleep in—like a clubhouse. Belok said he called it Dogs' Corner because he had lots of dogs and kept them there so they would not poop in the main village of Yakob. I found Dogs' Corner had a friendly feeling.

My most intimate fraternizing with Asabano canines was with a dog belonging to the family of a middle-aged spiritual healer named Walen. This female dog, who had been given the locally common dog name of Melise, adopted my wife Heather Miller and me, as well as the vacant house we shared during our 2005 stay in Yakob. Walen's wife Sansib told Heather and me that she had paid 100 Papua New Guinea kina for her—a substantial amount—but that she was a lazy dog who just wanted to sit around the house. Indeed, she frequented the comforts of our home. One day we found Melise waiting for us on our verandah. We were charmed and let her sleep inside. Heather described the event in more detail in her fieldnotes of that day, July 30, 2005:

> Breakfast. Warren's dog (Milise) is on our porch when we open the door, clearly expects to come in and have breakfast. We give her a bit of sweet potato, now friends for life (she even plays with us with her front paws). But we make her go outside when we eat. Unfortunately, she has pooped in the yard under the clothesline. One reason to discourage her from coming. [Laurie tells us later that she is a useless

Figure 8.6 A dog at Dog's Corner hoping for a bite of what Yalowad is eating in 2007. Photo: Roger Lohmann.

dog, can't hunt, just lies around the house and is lazy, and excretes near the houses—as we discovered ourselves! But we still like her.]

On August 13, Heather wrote, "Melise (Warin's dog) spent the night on our porch again. Roger gives her half a Wopa [brand] biscuit, and later some leftover sweet milky rice. She is VERY enthusiastic about sweet rice".

The next night there was a drum dance in the village and also heavy rain. Heather noted, "Melise is on our porch when we are escorted back by torchlight, spends night inside". And on August 15, "Melise is a bit over-excited today, has to be put outside during our breakfast—when we point and say 'lie down', she thinks we are holding out our hands to offer her food. Need to learn method of saying 'no' to dogs here. (Find out later it's sort of a hiss—SSST.)" On August 18, Heather recorded, "Melise is back this morning; again anxious for sweet rice—sugar dog, even licks up Milo [cocoa drink mix] powder when I spill some. We find out from Laurie when he comes for morning translation work that she is from Limbana".

Laurie's reference to Lembana probably indicates the more populous Omai settlement of Kusup, where I had noticed dogs behaving and being treated more like the pets of my own background. This background, Melise's difficulty fitting Asabano standards for how a good dog should behave, and her attraction to our household suggest the possibility that she had been previously socialized and enculturated into Omai standards before being sold away to a foreign culture. Melise was a misfit and therefore understandably unappreciated in the community in which she found herself, but in our home, she had a brief respite.

Similarly, Myrdene Anderson (2016) tells of being adopted by a dog of her Saami hosts. Although this reindeer herding dog, named Runne-Beane, was accomplished in the duties to which he had been assigned, Anderson (2016: 38) "noticed that Runne-Beane, his original mistress, and myself were all three considered to be cultural deviants: we exhibited our devotion to one another". As Ruth Benedict (1953[1934]) so forcefully demonstrated, the very same characteristic may make one valued in one ethnic culture and derided in another.

Melise's example indicates that dogs, like humans, are sensitive to cultural differences around them. Melise had been enculturated differently than the other dogs in Yakob, and because of this evidently preferred my family's foreign culture. In contrast, native Asabano dogs never approached me or tried to take up residence with me.

Ethnographer Yasmine Musharbash (2018:26–27) writes that her dog from Yuendumu, Central Australia

> was highly alert to the fact that I treated him differently to the way the other dogs in the camp were treated as well as from whom he could (and as importantly: could not) demand such different treatment: I offered pats, for example, and treats and he could insist on his prime seat in my car. Simultaneously, he endeavored to blend into the pack, making light of his different status. All this served him well when I had to move back to the city. I will never forget our first few visits to Fremantle Dog Beach … I could see him look at me quizzically, every so often, as he was trying to puzzle out the ways in which Dog Beach dogs communicated: so very differently from Yuendumu camp dogs!

Figure 8.7 Melise liked to warm herself by our hearth; Yakob village, 2005. Photo: Roger Lohmann.

> These dogs were much more keyed into their humans than camp dogs, and some were positively inept at dog communication. He made friends and settled in ... but it was only on the drive back to Yuendumu a year later that I realized how very much he had missed home.

Dogs, it would seem, can experience culture shock, and can learn a facility for transcending cultural differences, just as humans (at our best) can.

Conclusion

Dogs and humans are joint members in multispecies communities in which cultural representations of reality act on, and sometimes in, more species than just the humans who may fancy themselves above and beyond the rest. Dogs, like humans, must mentally represent cultural models of reality and propriety. I expect that the form, complexity, and function of their schemas is quite different from those of their humans to the extent that their brain-minds and the rest of their bodies are different. Dogs' life spans are much shorter than humans', they are better at certain things (like smells), and worse at others (like language). But both dogs and humans participate in the same communities and experience variable treatment depending on current and locally fashionable conventions.

Cultural differences that would not exist without the participation of multiple species are apparent in human–canine communities in New Guinea and elsewhere. As the ancient Roman poet Grattius wrote in his work on hunting, "Dogs have a thousand countries and each is in character what his country makes him" (quoted in Marvin 2005:61). This implies that both species share and produce a degree of the unique culture present in any society that includes or included dogs and humans.

References

Anderson, Myrdene. 2016. Runne-Beane: Dog herds ethnographer. *Ethnobiology Letters* 72(2): 32–40.

Anza-Burgess, Kasia, Dana Lepofsky, and Dongya Yang. 2020. "A part of the people": Human–dog relationships among the Northern Coast Salish of SW British Columbia. *Journal of Ethnobiology* 40(4): 434–450.

Bateson, Gregory, and Margaret Mead. 1942. *Balinese character: A photographic analysis.* New York: New York Academy of Sciences.

Beetz, Andrea, Kerstin Uvnäs-Moberg, Henri Julius, and Kurt Kotrschal. 2012. Psychosocial and psychophysiological effects of human–animal interactions: The possible role of oxytocin. *Frontiers in Psychology* 3: 234.

Benedict, Ruth. 1953 [1934]. *Patterns of culture.* New York: Mentor.

Bock, Philip K. 1999. *Rethinking psychological anthropology.* 2nd. ed. Prospect Heights, IL: Waveland.

Brightman, Robert. 1993. *Grateful prey: Rock Cree human–animal relationships.* Berkeley: University of California Press.

Brunois, Florence. 2005. Man or animal: who copies who?: Interspecific empathy and imitation among the Kasua of New Guinea. In *Animal names.* Alessandro Minelli, Gherardo Ortalli, and Glauco Sanga, eds. Pp. 369–381. Venice: Instituto Veneto di Scienze, Lettere ed Arti.

Canejo-Teixeira, Rute, Pedro Armelim Almiro, Luís V. Baptista, and Maria Manuela Grave Rodeia Espada Niza. 2020. Predicting dysfunctional human–dog dyads. *Anthrozoös* 33(6): 743–758.

de Vidas, Anath Ariel. 2002. A dog's life among the Teenek Indians (Mexico): Animals' participation in the classification of self and other. *Journal of the Royal Anthropological Institute* 8(3): 531–550.

Dwyer, Peter D., and Monica Minnegal. 2016. Wild dogs and village dogs in New Guinea: were they different? *Australian Mammology* 38: 1–11.

Dwyer, Peter D., and Monica Minnegal. 2021. Relationship between wild-living and village-living dogs in New Guinea. *PNAS* 118(12):e2020432118.

Fanaro, Luisa Amador. 2020. The domestic, the wild and its interstices: What can a dog do in Tierra del Fuego. *Vibrant* 17:a357.

Foley, William A. 1997. *Anthropological linguistics: An introduction.* Malden, MA: Blackwell.

Goodenough, Ward H. 1981. *Culture, language, and society.* Menlo Park, CA: Benjamin/Cummings.

Guillo, Dominique, and Nicolas Claidière. 2020. Do guide dogs have culture? The case of indirect social learning. *Humanities and Social Sciences Communications* 7(31): 1–9.

Haraway, Donna. 2003. *The companion species manifesto: Dogs, people and significant otherness.* Chicago: Prickly Paradigm.

Hecht, Julie. 2017. How much are dogs influenced by local culture? In *Scientific American*.

Herzog, Hal. 2011. *Some we love, some we hate, some we eat: Why it's so hard to think straight about animals.* New York: Harper Perennial.

Holland, Katrina E. 2021. Ambiguity, ambivalence, and affective encounters: An ethnographic account of medical detection dog–trainer relationships. *Anthrozoös* 35(3): 1–13.

Ivaşcu, Cosmin Marius, and Alina Biro. 2020. Coexistence through the ages: The role of native livestock guardian dogs and traditional ecological knowledge as key resources in conflict mitigation between pastorialists and large carnivores in the Romanian Carpathians. *Journal of Ethnobiology* 40(4): 465–482.

Jorgensen, Dan. 1980. What's in a name: The meaning of meaninglessness in Telefolmin. *Ethos* 8(4): 349–366.

Jorgensen, Dan. 1981. Taro and arrows: Order, entropy and religion among the Telefolmin. Ph.D. dissertation, Anthropology, University of British Columbia.

Keil, Paul G. 2015. Human–sheepdog distributed cognitive systems: An analysis of interspecies cognitive scaffolding in a sheepdog trial. *Journal of Cognition and Culture* 15(5): 508–529.

Kohl, Johann Georg. 1985 [1860]. *Kitchi-Gami: Life among the Lake Superior Ojibway.* Translated by L. Wraxall. St. Paul, MN: Minnesota Historical Society Press.

Kohn, Eduardo. 2007. How dogs dream: Amazonian natures and the politics of transspecies engagement. *American Ethnologist* 34(1): 3–24.

Kondo, Shiaki. 2021. Dog and human from Raven's perspective: An interpretation of raven myths of Alaskan Athabascans. *Polar Science* 28: 100633.

Lambert, Joanna E. 2012. Primates in communities: The ecology of competitive, predatory, parasitic, and mutualistic interactions between primates and other species. *Nature Education Knowledge* 3(10): 85.

Lohmann, Roger Ivar. 2004. Culture. In *Encyclopedia of religion*. Lindsay Jones, ed. Pp. 2086–2090. New York: Macmillan Reference.

Lohmann, Roger Ivar. 2011. Empathetic perception and imagination among the Asabano: lessons for anthropology. In *The anthropology of empathy: Experiencing the lives of others in Pacific societies.* Douglas Hollan and C. Jason Throop, eds. Pp. 95–116. New York: Berghahn.

Luomala, Katharine. 1960. The native dog in the Polynesian system of values. In *Culture in history: Essays in honor of Paul Radin*. Stanley Diamond, ed. Pp. 190–240. New York: Columbia University Press.

Marvin, Garry. 2005. Disciplined affections: The making of an English pack of foxhounds. In *Animals in person: Cultural perspectives on animal–human Intimacy.* John Knight, ed. Pp. 61–77. Oxford: Berg.

McCormack, Patricia A. 2018. An ethnohistory of dogs in the Mackenzie Basin (western Subarctic). In *Dogs in the north: Stories of cooperation and co-domestication*. Robert J. Losey, Robert P. Wishart, and Jan Peter Laurens Loovers, eds. Pp. 105–151. London: Routledge.

Mezzenzana, Francesca. 2023. Just like humans: Similarity, difference and empathy towards nonhumans in the Amazonian rainforest. In *Conversations on empathy: Interdisciplinary perspectives on imagination and radical othering*. Francesca Mezzenzana and Daniela Peluso, eds. Pp. 197–215. London: Routledge.

Mittelbach, Gary G. 2011. *Community ecology.* Sunderland, MA: Sinauer Associates.

Musharbash, Yasmine. 2017. Telling Warlpiri dog stories. *Anthropological Forum* 27(2): 95–113.

Musharbash, Yasmine. 2018. Yuendumu dog tales. In *Living with animals: Bonds across species.* Natalie Porter, ed. Pp. 17–28. Ithaca, NY: Cornell University Press.

Payne, Elyssa, Pauleen C. Bennett, and Paul D. McGreevy. 2015. Current perspectives on attachment and bonding in the dog–human dyad. *Psychology Research and Behavior Management* 8: 71–79.

Plata, Elías, and Salvador Montiel. 2020. Human–dog bond in the contemporary Mayab: Social perceptions and benefits associated with the hunter-Milpa dog in Maya peasant-hunter life strategies in Yucatan, Mexico. *Journal of Ethnobiology* 40(4): 451–464.

Rose, Deborah Bird. 1992. *Dingo makes us human: Life and land in an Australian Aboriginal culture.* Cambridge: Cambridge University Press.

Steadman, Lyle B. 1971. Neighbours and killers: Residence and dominance among the Hewa of New Guinea. Ph.D. thesis, Australian National University.

Tempelmann, Sebastian, Juliane Kaminski, and Michael Tomasello. 2014. Do domestic dogs learn words based on humans' referential behavior? *PLoS One* 9 (3):e91014.

Titcomb, Margaret. 1969. *Dog and man in the ancient Pacific with special attention to Hawaii.* Honolulu: Bernice P. Bishop Museum.

Willerslev, Rane. 2007. *Soul hunters: Hunting, animism, and personhood among the Siberian Yukaghirs.* Berkeley: University of California Press.

Williams, F. E. 1936. *Papuans of the Trans-Fly.* Oxford: Oxford University Press.

Wynne, Clive D. L. 2019. *Dog is love: Why and how your dog loves you.* Boston: Houghton Mifflin Harcourt.

9 Going to the Dogs with Enewetak/ Ujelang Marshall Islanders

Laurence Marshall Carucci

In a classic article from the 1980s, James Clifford discusses the integrally inscribed manner in which the authorial voice is unavoidably woven into all ethnographic accounts (Clifford 1983). Given this truism, as a lifetime lover of dogs who has spent endless hours accompanied by intelligent canines, my ability to judge the interrelationships among dogs and humans in the Marshall Islands immediately places issues of bias on center stage. Far more than other topics, given my own feelings of dis-ease when watching Marshall Islander's dogs be "abused" by humans (in American terms), this chapter undoubtedly over-represents my own biases, even as I pen a story that attempts to accurately interpret the intersubjective, lived relationships I have experienced while "deep(ly) hanging-out" with, working alongside, and observing the interactions of hundreds of Marshall Islanders during the past five decades. Quite clearly, as Enewetak/Ujelang people have become more tightly integrated into a world capitalist system, their own interactions with dogs have changed, much as, in several ways, they have remained the same. This shifting dynamic is obvious given more than two dozen research stints I have spent with members of the Ujelang/ Enewetak community and other Marshall Island's communities in the Republic of the Marshall Islands and beyond.

In this chapter, I track the contour of these changes alongside the facets of canine–human relationships that have remained largely unchanged. Not surprisingly, these shifts in interaction provide a condensed example of larger changes and continuities in Marshall Islanders' culture and everyday activities. While investigating the contours of the interspecies relationship that, for Marshall Islanders, interweaves the lives of people and their dogs, I shall introduce a moment of pause and serious introspection into the contention that dogs are culture-bearing species (Lohmann, this volume, Chapter 1). There is little doubt that dogs are thoroughly culturally-constructed critters and their complex social relationships with humans is undeniable. Equally, dogs, like all living things, are semiotic creatures. Dogs use their sign-based faculties, grounded in the manipulation of icons and indexes (Peirce 1960 [1931]:2.230 *et seq.*), in very complex ways to interact with other living beings and humans as they live their lives in the world.

DOI: 10.4324/9781003658696-9

Nevertheless, even though dogs and humans certainly relate to each other as part of an interspecies social order that involves meaningful and dynamic interactions for individual members of each species, I do not believe dogs possess cultural abilities grounded in the manipulation of symbols (Kohn 2013: chapter 4 and pp. 222–223). As Dobrin (this volume, Chapter 2: 55) suggests for the Arapesh, dogs and humans utilize entirely different communicative modalities. As she says, "Where there is no shared language, speaking must take place through signs and gestures." But, it is not as though such signs and gestures lack the ability to convey meaning. For both dogs and humans, they do. But, seeking the varied forms of this perspectivism or multinaturalism, as Viveiros de Castro terms it (2004; 2016), it is critical to remain true to the anthropological project of seeking a legitimate understanding of radical alterity. Anthropologists have long relied on long-term interactive research to attempt to comprehend the perspectives of others in their own terms and according to their own rules. Extending this same method to understand the radical alterity of other species, even, perhaps, so-called inanimate entities like rocks (Povinelli 1995), does not mean extending "culture", qua human's symbolic facilities, to dogs or other beings. Such an approach provides only an anthropocentric mirage of such multinaturalist relationships (Viveiros de Castro 2016: 244–245). Instead, cautiously translating the multilayered significata of the signs and gestures (iconic and indexical representations) used by both dogs and humans across species boundaries, extends the anthropological methods used to comprehend radical alterity into a new domain in a careful manner so that translations, always the core piece of constructing intersubjective "others" in the rendering of radical alterities, retain the authorized traces of their representational method. The semiotic dimensions of dog–human interactions are explored in greater depth below to attend to the complex universes of dog–human interaction strategies and actions, keeping in mind the ways that both humans and dogs rely on a panoply of indexical and iconic modes of representing and interacting in their world. These multi-species interactions between dogs and their Marshall Islands companions, both extend beyond culture and, for humans, evoke the residual traces of an ancient, pre-symbolic, but densely semiotic, hominid past. This past continues to have resonance for contemporary humans in their complex, multi-layered communication strategies with one another and with other entities, both living, inanimate (or very slowly shifting in shape), or non-corporeal in their contours.

The Shifting Contours of Canine–Human Relationships among Marshall Islanders

My sense of culture shock, the dizzying feeling of unsettlement and anomie, was at its maximum when, in 1978, I returned from living in the Marshall Islands for more than two years to Hawai'i, and then to the mainland United States. Nevertheless, when I initially arrived in the Marshall Islands in 1976, I was aghast at the treatment of dogs in Majuro, government center of the Marshall Islands. In my first few days on Majuro, the frequent ill-treatment and general state of impaired health of Majuro dogs was repulsive. Operating at that time on the American trope that dogs were human-like, "man's best friend", or perhaps, even better friends for many than their hominid consociates, it concerned me that Marshall Islands' dogs were so mistreated. Of course, comparable levels of animal mistreatment are found elsewhere as Morrison notes in her contribution to this volume. Clearly, Americans' love of dogs is also tinged with ambivalence. As Morrison notes, while abandonment and lack of pet care set the conditions for euthanizing animals in Kea'au, the anxiety and stress created among animal shelter workers when dogs must be euthanized certainly derives from the friendship and consideration that are grounded in the standard household positioning of dogs as loved beings worthy of indulgence and care in most sectors of Hawaiian society, as on the mainland United States.

The mistreatment of dogs in Majuro caused me anxiety that kept me awake at night. I wondered how Marshall Islanders might treat each other given their unsettling treatment of dogs. Owners of the hotel where I stayed upon first landing in the Marshall Islands kept a truly mangy dog. While he was fed and not mistreated, his state of ill health worried me. Other Majuro dogs that I encountered in those introductory days differed only slightly. Some, while hardly groomed, had less mange, but often sported wounds caused by coral rocks flung their way as they raced into the road to greet passers-by with threatening lunges and growls. These canine encounters occurred prior to the time I first met members of the Ujelang/Enewetak community with whom I would spend a good part of my life.

Arriving on Ujelang a couple of weeks later, no canine companions accompanied the swell of community members who came to the shore to formally shake my hand in greeting. Within days, I was told that all the dogs had been killed and eaten nearly a decade earlier as part of a prevention strategy to rid the atoll of an infestation of "white people's" rats (Norwegian rats). Indeed, at that time, Ujelang law dictated that no dogs were permitted. Public health representatives from the Marshall Islands had insisted local people get rid of all dogs so that cats could rule the land and rid the main islet, and a few of the other large, habitable, islets, of the infestation of rats. Importantly, Ujelang people themselves were ambivalent about eliminating the dogs that shared their communal space, in some cases, for years, but state authorities gave the community no choice. They insisted the dogs be killed to maintain public health. Such an eradication of dogs also occurred among the Inuit of Cumberland Sound, with the same rationale of protecting public health but, as Lévesque notes, colonial strategies and practices varied, depending on timing and circumstances. In other Inuit locations, at other historical moments, very different rules applied (Lévesque 2018:184–185). Similarly, on Ujelang the long-standing inter-species relationships between local people and their dogs did not die with the state-imposed "no dog" rule. Historically, *kidu* "dog(s)" have been part of Marshall Islander's lives at numerous times since these tiny spits of sand became habitable, shifting in people's imagination between consociate species and sustenance (Cramb 2021:30–31).

Indeed, a few months after my landing, the grandson of one high chief returned to the atoll with a dog. In line with the law, the dog was not permitted to reside on the main islet of Ujelang, but the Council decided that no one could prevent the young man from keeping the dog on Bieto, a neighboring islet that belonged to his grandfather, Chief Ioanej. The "no dog" rule was interpreted (or re-interpreted) to mean, at least in the aftermath of the rat infestation, that dogs only could not live on Ujelang islet of Ujelang Atoll, the main residence islet. The importer's chiefly status was said, by some, to have influenced the easing of the dog-ban. Within months of the first dog's arrival, another pregnant dog was placed on Bieto. Lacking any full-time residents, those feral dogs are said to have foraged on the reef at low tide for fish and sea cucumbers, and to have eaten lots of coconut. While they were hardly well-fed, being small in number, they survived. Their diet made them particularly greasy, or so I was told by young men who killed the maturing young puppies, and consumed them during drinking fêtes that took place on Bieto or on the leeward end of Ujelang islet (Carucci 1987). Nearly a year later, participating in one such drinking circle of young men, I sampled my first taste of roasted dog, a cut from the hind-quarter of the mother dog who had birthed the puppies that, according to other young men attending the event, were said to be more delicious than the mature canine we consumed.

While I refer to these Ujelang dogs as "feral" in line with the usage of the term by Koungoulos (this volume, Chapter 7), Ujelang people term them *awia* "wild", since they do not inhabit the domestic space with co-resident humans. Certainly, they are dogs of European ancestry, not descendants of ancient Marshall Islands' dogs brought by the earliest Marshall Islands voyagers (see Weisler 2001 for evidence of these pre-contact dogs on nearby Utedik). Nonetheless, other than the mode of slaughtering them, these Bieto dogs, unlike those on Majuro, suffered no human abuse. At their moment of demise, Ujelang/Enewetak dogs that are killed are felled with sizable coral rocks hurled by young men, then dispatched with a machete by slitting the throat of the animal. Indeed, this is exactly the same method used when capturing pigs for slaughter, the other four-legged food item in the Ujelang/Enewetak array of edibles.

Given my love of dogs, consuming several bites of the mother dog during a drinking fête was, at best, an ambivalent experience (cf. Haines and Sammels 2010). Nevertheless, as a young anthropologist following the proscription, "do as local folks do", my choices were limited. This was particularly true given the rules of young men's drinking circles where each participant consumes "identical" amounts of drink—in this instance *iiej* "yeast" (a noxious local brew concocted from baking yeast, sugar, and sprouted coconut)—along with available foods (Carucci 1987). My fellow drinking circle participants, of course, saw the sharing of drink and food as tests of my solidarity with members of the community and their drinking group. Refusing to participate would not have enhanced my ability to work with the maturing young Ujelang men.

Moving to the New Enewetak

Ujelang people's decision to eliminate their population of dogs in the 1960s was a direct outcome of their experiences of living in exile while the United States conducted nuclear tests on their home atoll, Enewetak. Known to military veterans as "Eniwetok", the atoll had been a major 1944 battle location as the Allies moved persistently across the northern Pacific on the "Second Road to Tokyo" (Poyer, Falgout, and Carucci 2000). Over 20% of Enewetak people were killed during the Battle of Eniwetok, leaving fewer than 140 living residents. For those survivors of the battle, WWII did not end on VJ Day, August 14, 1945, as U. S. military occupation of Eniwetok continued. Kwajalein and Eniwetok simply shifted their mission, becoming support facilities for the Pacific Proving Grounds. The first Crossroads atomic explosion, a test dubbed "Able", took place on nearby Bikini on July 1, 1946. By December of 1947, Enewetak people were moved from the fringes of their home atoll to Ujelang as nuclear testing was expanded onto the northern islets of Eniwetok at the same time tests continued on Bikini. While Enewetak people were promised that they would soon be returned to their homeland, they spent the next 33 years living in exile on Ujelang. As they transformed themselves into diUjelang, literally, the people of the Middle of Heaven, life on Ujelang proved less than heavenly. On this small atoll, 1/14th the size of Enewetak, strewn with coral rubble from an 1860s typhoon, people suffered through sequential waves of famine as the community was left without supply ships carrying supplementary market commodities for stints of four to eight months. During the rat infestation, small coffins were designed to protect infants' fingers, toes, and noses from being chewed on as they attempted to sleep through the night. This was only one part of the suffering endured by local people during their years of life on Ujelang.

Dog rules changed when Ujelang residents were repatriated to Enewetak, their home atoll, in 1980. When I lived on Enewetak in 1982–83, dogs were a common sight; their treatment aligned with historical experience, and what people had witnessed when they visited Majuro. One difference, however, was that Enewetak dogs were eaten, whereas Marshall Islanders from Rālik and Ratak prohibited the consumption of canine flesh. Marshall Islanders long have considered Ujelang/Enewetak people to be inferior, outback "hicks" of the Marshalls, and eating dog in the late 20th century, was one core marker of their low rank. However, for Ujelang residents, who had struggled with waves of starvation during the 1950s and 1960s (Carucci 1980; 2011), dog was a perfectly acceptable food. Some claimed that Enewetak and Ujelang people had eaten dog since time immemorial. Others contended, they learned how to prepare delicious roasted dog from neighboring folks from Pohnpei, Mōkil, and Pingelap. Contact with these Eastern Caroline Islanders was frequent during the Japanese era prior to World War II when Ujelang and Enewetak had been governed from Pohnpei, not from Jaluij or Majuro.

Jaluij, governmental center of the Marshall Islands during the German era (1885–1915), continued as the Marshall Islands capitol during Japanese times for all Marshallese speakers other than Enewetak and Ujelang folks. However, once the United States established governmental control of the Trust Territory of the Pacific Islands during the Navy's administration of the area following WWII, Jaluij was replaced by Majuro as the government center, with Ujelang and Enewetak once again part of the Marshalls. Majuro remained the administrative center throughout the Trust Territory of the Pacific Islands era, and continues as the capitol of the Republic of the Marshall Islands today. For Enewetak/Ujelang people, part of their marginal identity has been dependent on their shifting inside-versus-outside position living, as they do in the borderlands, in "the middle of the ocean", simultaneously on the margins of the Ratak/Rālik (Marshall Islands') seascape and the Pohnpei/Eastern Carolines' seascape.

Dogs and Drudgery on Enewetak

One huge problem for dogs, pigs, and other animals on Enewetak, was that there was no food for them to eat.[1] When Enewetak/Ujelang people were repatriated on Enewetak in 1980 following a radiological clean-up of five southern islets of the atoll, the United States placed them on a reservation-style diet consisting of USDA commodities and other supplemental foods. Since all vegetation and topsoil had been stripped from the rehabilitated islets, it would be decades before any food-producing trees could be grown successfully in the depleted and toxin-laden soil. Unfortunately, the rehabilitation plan included no food for pigs or chickens, much less for dogs and cats. All of these animals were fed left-over food scraps as well as some canned vegetables that local people themselves struggled to eat since, as they said, they "did not know anything about their scents/flavors". Lacking dedicated supplies of animal food, the squeals of hungry pigs on this "new", environmentally-hollow, Enewetak were incessant during the 1980s and 1990s. Having expected to return to the verdant atoll they remembered from pre-war years, people found it impossible to live in the Marshallese style (subsisting on foods from the land and sea) on the fragmented remains of their one-time homeland. Within two years of their return, local residents began calling their grossly-transformed "homeland" the "New Enewetak". Lacking local foods to forage and scavenge, dogs inevitably sought out human food-sharing events. In return, canines were rewarded with a few morsels of discarded food and multiple volleys of coral rocks. A few dogs were more indulged than others, receiving cast-off food with larger edible residues and fewer warning hisses, but, while valued for their barks of early warning, none were treated as pets. Indeed, even after years of living as part of the Enewetak/Ujelang community, I always cringed when young children were allowed to carry young puppies by a single leg, swinging them about as they played drowning games with young dogs or cats, and when they treated them in an antithetical manner to what *di pālle* "clothed white people/Americans" would call humane.

Of course, "humane" has its own shifting etymological history in English, reflecting larger shifts in British and American worldviews. It now refers to that which is human or human-like, but derives from an earlier era when the term referred to earthly beings as distinguished from gods. The current sense of humane arises in the 1600s with its focus on compassion and kind treatment of others, differentiating it from earlier uses that designated civility, friendly demeanor, and obliging practices in a person's interactions with others (Etymological Dictionary online). The newer compassion-focused meanings, are clearly the ones now extended to animals. Those characteristics underlie referents like The Humane Society, founded in 1954.

Not unlike the treatment of puppies among Yapa (Walpiri people) in central Australia, the treatment of dogs, and of children among Enewetak/Ujelang Marshall Islanders—including the freedom from negative sanction—tells much about the ethical and moral constructions of adult members of the community (Musharbash 2017). While the ill-treatment of dogs by adult Enewetak/Ujelang people was certainly not encouraged, neither was it negatively sanctioned. Not surprisingly, then, much as the treatment of dogs in Majuro provided the encounter that, for me, most closely approached culture shock, the ill-treatment of animals, particularly their abuse by young Marshall Islands' children, framed the moments when I have felt least like a member of the Enewetak/Ujelang community.

In this volume, Prokosch (Chapter 3) describes dog–human interactions in Majuro. His depiction of the treatment of puppies and dogs by young children aligns with my own unsettling observations and understandings, not only in Majuro, but in other locales in the Marshall Islands. While my experiences of humans' interactions with dogs in Majuro suggests a much more diverse set of interspecies relationships between dogs and their Majuro owners than those described by Prokosch, his account certainly covers one dimension of those canine–hominid interactions. Not infrequently, dogs in Djarrit/Uliga/Delap (the urban corridor of Majuro designated DUD) are encouraged to be far more aggressive than those in the increasingly rural corridor that extends from Rairek to Majuro islet (Laura). Other exceptions to Prokosch's depictions include canine pets owned by Americans or certain other foreign nationals who reside in the Marshalls.

Dog Marginalization

Relying on functionalist rationales, several participants in the "dogs session" at the 2021 ASAO Meetings, suggested that the lack of dogs dwelling with humans who reside on Pacific atolls was due to the minimal need for their services as hunters, trackers, pullers of heavy loads on sleds or sledges, etc., that is, the types of activities that working dogs among Utku, Inuit, or Hare residents of America's northlands have found extremely useful (Savishinsky 1975). While fine-grained human understandings of the limited land spaces on atolls may minimize the need for dogs as working animals, it seems risky to conclude that the functional contributions of dogs to the productive life of humans exhaust their potential values. Indeed, the historical shifts in Enewetak/Ujelang people's relationships with dogs, and the varied positioning of dogs in Feinberg's and Pyrek's experiences on other Pacific atolls (this volume, Chapter 6), suggests that more nuanced interpretational frameworks are required. Dogs and cats in contemporary European and American societies have symbolic and psycho-social values that frequently far exceed their value in everyday productive activities (Sahlins 1976: 170 et seq.; Haraway 2003). A comparable array of complexities in human/canine interactions should be expected on Pacific Islands atolls.

A less *Homo*-centric, interspecies approach, allows us to better assess these complexities. Such an interspecies view necessitates the assessment of the dog's perspective alongside that of the humans with whom they live and interact. Research by Perri et al. (2021) suggests that, far from humans making dogs from wolves for their own ends, domestication was not a homocentric activity, but rather represents an "emerging mutual relationship" between the two species. This dog-human relationship, the only domestication event to have occurred in the Late Pleistocene, was likely dependent not only on climatic shifts, but on increased interactions "perhaps resulting from the mutual scavenging of kills, or from wolves drawn to the detritus of human campsites" (Perri et al. 2021:6). Wishart (2018:1) agrees with this mutually beneficial perspective, noting that the co-domestication of wolves in interaction with humans probably occurred many times, either through "copying" or through "independent realization of positive inter-species reciprocity". Advantages to both species caused wolves and humans to move into close interactive relationships with one another, creating conditions of co-domestication (see also Hare et al. 2002; Shipman 2015). Following Haraway (2008), Wishart notes that "without dogs there would be no humanity as we know it". Perhaps this was most true in the north (Wishart 2018:2).

Correlating human genetic ancestry with dog haplogroup evidence, Perri et al. believe that Ancient North Siberians were the most "likely population to have initiated the domestication process", allowing for "the transfer of dogs into different groups and thus their movement both east and west following their domestication" (Perri et al. 2021:6). Wishart notes that the benefit to early canines included a more stable food supply along with protection, while the benefits to humans came in the forms of tracking and hunting as well as protection (Wishart 2018: chapter 1). Iditarod fans may wonder about the renowned sled-pulling capabilities of northern breeds, but McCormack demonstrates that dogs shifted from being hunting companions to sled-pulling athletes during the fur-trading era in Canada (McCormack 2018: chapter 7). This was not the first dynamic shift in the co-domestication of dogs and humans. In the same volume, Hill, relying on archaeological evidence, shows that human–dog relations shifted substantially in Northwest Alaska from about 700–1,000 C.E. from cohabitation to co-dependency, a move that increased the likelihood of survival for both species in this particularly marginal environment. The shift is marked by dogs becoming meaningful members of families but, also by dogs being used as food, or enduring suffering death as sacrificial stand-ins for humans, indicating that dogs and humans were both thought to possess something like a spirit essence (Hill 2018: chapter 6).

At the same time, the archaeological record provides Hill with evidences of other ambivalences and ambiguities in the shifting co-dependence of Northwest Alaskan social practices with the killing and eating of dogs, at least during times of famine. She also finds signs of dog abuse in the archaeological record. Therefore, alongside the benefits to humans provided by their dogs in terms of hunting and tracking, transportation, trade, and travel, keeping dogs also required setting aside extra resources, simultaneously a mark of their value and of their cost (Hill 2018: chapter 6). The same ambivalences can be seen on Ujelang and Enewetak, where dogs were highly valued, yet, in certain times and particular circumstances, they were abused and consumed as food. Certainly, prior to the time dogs were reintroduced by the chief's grandson, local people nostalgically discussed the fact that they missed having dogs as members of the community. Yet, a few community members also noted that during the waves of famine Ujelang residents began to endure around 1952–53 on Ujelang, resident dogs were eaten. When times of famine subsided, dogs were reimported as community cohabitants. In all likelihood, this was true on other Marshall Islands atolls in pre-European times, before foods from the world market became available. Nowadays, however, with the eating of dog representationally tinged by its inscriptions of primitive as opposed to civilized choice, the most empowered members of the Enewetak/Ujelang community have only rarely consumed dog. By contrast, young men (and a few young women), who may gain renown through challenging the contested norms of everyday life, engage in the practice of eating oven-roasted dog with some regularity.

The most noted advantage of canines over hominids, lies in their superior capabilities of smell. Tracking and hunting are derivatives of that ability. In her research on dogs, Horowitz stresses that dogs live in worlds in which they constantly map the shifting contours of scent, not simply "reading the news" embedded in scents deposited on apocryphal fire hydrants, but interpreting the ever-shifting contours of scent carried by the wind, and the archaeological record of scents distributed through the earth, sea, and sky (Horowitz 2016). She contends that popular articles whose authors make claims of "dog's noses being 10,000 times more sensitive than humans" are conflating a nuanced panoply of physical enablements that underlie dogs' superior scent sensibilities. While mammalian brain processing of scents is accomplished in a similar structural manner, the differential detection parameters separating dogs and humans are inappropriately reduced to a single dimension by simple multipliers of human capabilities. Human cultural sensitivities and classifications of scent further complicate any such universalistic comparisons.

While Horowitz' contention that dogs' scent detection capabilities far exceed that of humans is undoubtedly true, she frequently generalizes too readily to all types of dogs. By contrast, Coren, a long-time dog breeder, competition judge, and researcher, argues that breeds of dogs differ significantly from one another in terms of their abilities, including abilities dependent on scent discrimination (Coren 1994). Not surprisingly, Coon Hounds and Siberian Huskies are far more successful hunting and tracking dogs than Chihuahuas. But, as acute as husky and hound noses may be, they are far more interested in consuming or hounding other animals than in herding, a special skill of Australian Shepherds.

What is true of all these canine-specialized, but breed-specific characteristics, is that, while "culture-like" in their distinct form that differs from breed to breed, they are genetically-inherited capabilities that are then environmentally contoured. While all contributors to this volume provide a plethora of materials that allow us to explore the interspecies interactions of humans and dogs in the Pacific, I believe it is essential to critically examine the use of the term "culture" to describe the conceptual and communicative characteristics of dogs in their interactions with humans. Certainly, Lohmann (this volume, Chapter 8) asks that we utilize the concept of interspecies culture to discuss these interactions, and he utilizes the work of Guillo and Cladière (2020) to decide if particular non-human animals are acting in a cultural manner. Such an approach, however, massively reduces the meaning of culture in the sense that it has been understood and refined by anthropologists since the time of Boas (Stocking 1968: chapter 9). In accord with that anthropological definition, the use of a conventional, learned, code of arbitrary symbols to communicate with one's cultural consociates, lies at the core of humans' cultural abilities. In Geertz' often-cited iteration: (each human culture entails) "a system of inherited conceptions expressed in symbolic forms by means of which men communicate, perpetuate, and develop their knowledge about and attitudes toward life" (Geertz 1973). While anthropologists have moved beyond Geertz, incorporating a focus on social practices and contexts of interactions, it would be a huge oversight not to recognize that the learned, symbolic communications modalities discussed by Geertz, are the key operational concept of Boas' "cultures" (plural) constructions, lying at the core of human lives and differentiating humans from other species. No matter how complex the learning strategies of a Chihuahua may be, it will not become the equivalent of the hound or northern-breed hunter and tracker. By contrast, as Boas discussed, a person of Chinese ancestry raised from birth in New York will be a New Yorker. That person will not know Chinese nor interpret the world in a Shenzhen or Beijing manner unless specifically taught to be bicultural. Such is the potency of cultural sensibilities, and minimalist views of culture, like that of Guillo and Cladière, can lead us to lose track of the potency of culture writ large.[2] While dogs are constructed as cultural critters to their very core by humans, while dogs clearly represent humans in very complex semiotic ways using iconic and indexical representational relationships (Peirce 1960 [1931]:2.230 *et seq.*), making this interspecies communication of mutual benefit to the ways humans and dogs shape their own lives, this does not make dogs cultural, symbol-using creatures, of the same sort as their human companions. As Kohn stresses, a "multispecies ecology of selves … depends for its existence on the continuous flourishing of dense nonhuman ecologies … (and) on the humans who live by tapping into those ecologies" (2013:227). Kohn's multispecies ecology is larger than culture (2013: introduction and conclusion) for while humans and other living creatures, including dogs, actively communicate using deeply meaningful systems of signification relying on indexical and iconic signs, it is humans who are symbol-using beings, and that symbolic facility lies at the core of culture. Both dogs and humans are inherently social beings (Wishart et al. 2018: chapter 1), and both rely on sophisticated semiotic strategies for communication, but extending the term "culture" to dogs, in my estimate, only serves to perpetuate misunderstanding as we explore the type of multispecies ecology that allows for a larger-than-cultural, decolonized anthropological understanding of the world (Kohn 2013; Viveiros de Castro 1998).

Even though anthropologists are well aware of humans' cultural advantages and perhaps, equally, their potential constraints, such is not the case for scholars from other disciplines. Even my closest history colleagues in Montana frequently use "culture" to mean something very different than do I. Not only are there huge adaptive advantages of cultural innovation in relation to the slow pace of evolutionary change; equally, the ability to imagine and communicate about the universes of Narnia, Star Trek, or Harry Potter stand apart from the capabilities of other socially-sophisticated semiotic creatures.[3] But, if cultural production is prodigious and of extraordinary value, it also carries certain risks. That risk, one of genre confusion, of misapprehending the alignment or misalignment of symbolically-coded messages with the indexically-designated facticity of a "real" lived world, is a uniquely human risk.

The same quality that allows humans to imagine a scene in Narnia or Harry Potter, and certainly for actors to enact it qua performance, also allows some human actors to believe that ideological messages not grounded in lived reality are "in fact" true (not only in one's cultural imagination). In other terms, those actors believe in an all-encompassing ideological claim at the same time they fail to believe that the experiential realities of everyday life are, in fact, true. For U. S. residents, the representationally contested ground around the 2020 election and the storming of the U.S. Capitol, by "rioters and insurrectionists" or by "patriotic supporters of the 45th President" serves as a good example. Perhaps even more didactic cases may be found among today's array of climate change deniers, or those who ingested Ivermectin rather than a Pfizer, Moderna, Johnson & Johnson, or AstraZenica-developed vaccine as a preventative for serious SARS-CoV-2 illness. Lohmann (personal communication) correctly asserts that ideologically-grounded interpretations that are not aligned with the facts on the ground, have the same character as religious beliefs in general. Beliefs of this sort create solidarity among their believers, but, since they are grounded in faith, they need not be cross-checked or tested as verifiable facts. Without such cross-checking, however, symbol using humans are at high risk of misinterpretation since the signifiers "represent via their relation to an absent system of other such symbols that make them meaningful" (Kohn 2013:212). By contrast, indexical and iconic representations necessarily retain a connection between their signifiers and that which they represent (Peirce 1960 [1931]), not allowing space for deceit. While constrained by earlier-imagined potentialities and approaches of anthropology, Jules Henry worried considerably about the contorted effects of Americans' fears and fabrications in *Culture Against Man* (Henry 1963). At that time, Americans were focused on us-versus-them fears vis-à-vis the Soviet Union. Nowadays, Russia is only one of several newly-bedecked imagined others, both external and within national borders, for U. S. residents to fashion and fear.

Eduardo Kohn (2013: chapter 6), following Viveiros de Castro (1998), seeks to provide a post-colonial way to imagine human relationships to non-corporeal beings, one not constrained by the dyadic contrasts of culture versus nature or the real versus the supernatural, that sphere beyond the mundane. In Kohn's approach, it is necessary to consider absences since "all signs involve a relationship to something not present", be they icons, indexes, or symbols, and also because life itself, "being intrinsically semiotic, has a related association to absence" (Kohn 2013:212). These absences are all highly constitutive. Therefore, for jaguars, much like Siberian Huskies, to pounce on prey, the signifier, the image of the movement of an animal of prey, is displaced, moved "forward" in time and space (into a not-yet-present sphere, but one grounded in immediate past observation), to allow the jaguar to spring and capture its prey with considerable accuracy. Such is the highly productive capacity of indexical signifiers. Quite frequently, I have watched my Siberian Huskies in numerous such four-legged fox-like grand jeté, as they adjusted their flight mid-air before pouncing on prey to account for the slightly different trajectory that the prey animal actually took as it attempted to sprint to safety. But Kohn, following Viveiros de Castro (1998) notes that for humans, this imagining of the absence, involves constructing a space for the *you*, not only the invisible part of the objectified "self" (itself a secondness) (Kohn 2013:211), but a realm of otherness that is not just about the *I*. For Viveiros de Castro (1998:483; 2016:244–245), it is the *you*, the second person, that occupies the missing space between the "*I*" of culture (the person's view of her own self) and the impersonal "*it*" of nature.[4] This "other as subject … a latent echo of the '*I*,'" tells us something important about the positioning of the supernatural, a "realm", adds Kohn, "that is not just reducible to nature, nor … to culture" (Kohn 2013:214). It is this culturally-imagined, yet absolutely real, higher-ordered other positioning of the self, "both strange and familiar", where both the superhuman and the other-than-human find their logical representational, and deeply felt, position. But, it is easy for misapprehension to overwhelm this space. As Kohn notes: "The challenge is to avoid becoming an object in the process of this interpellation. And this is a real danger" (Kohn 2013:214). It is in this sort of misplaced concreteness, this **real** danger, that the misleading imagined components of ideology become experienced as the real by humans. And semiotic confusion of this sort is something dogs will never share.

The arbitrary nature of symbolic representation, that is, the absolutely non-linked relationship between signifiers and signifieds both allows for incredible cultural productivity and for the ready replication of propaganda. Such is not true of canine communication.[5] As noted by Kull, the representational potentialities of other species, including dogs, while densely semiotic, differ in a critical way from human communication modalities (Kull 2008:3212). In the sophisticated mapping of scent described by Horowitz (2016), dogs indexically map all of the shifting currents of scent and historically classify the presence of scent on objects in a manner that allows them to accurately reconstruct the extant condition of those elements in their environment. Even though each linguistic signifier used by humans in any language has several indexical layers of meaning attached to it (Silverstein 1992), most designata in spoken languages also traffic in a wide array of types as representations of their associated tokens. As symbols, no necessary relationship links signifier and signified. Even though the productive advantages of such arbitrary human linguistic signifiers are substantial, such arbitrariness also opens up possibilities for massive misrepresentation. Therefore, the latent scent of a dead deer carcass signals to a dog, a dead deer, when and where it died, what sorts of processes led to its putrification, and other components of that particular deer's identity and station in life/death. A learned human, enculturated in deer ecology, might well come to similar conclusions even if s/he could not interpret the scents as readily. However, a racist human interpretant might rather attribute the stench to the presence of new neighbors, people from Mexico, Afghanistan, or another "othered" locale, who may well be imagined not to share the interpretant's own particular cultural sensibilities regarding proper hygiene. A dog's indexically-grounded consciousness could never directly entertain such misalignments in attributing meaning to "the same" scents. That is, without pointing to the ideology of attributed otherness, indexical understandings necessarily remain attached to current-state-of-the-world contours regarding the extant conditions of scent. This is necessarily true because the signifier, the deer's scent, bears an indisputable link to the signified, the former or current presence of that deer. Given these differential semiologies of scent, the dog will seldom be misled by his/her mapping of the scent-scape. A dog will sniff around, following historically older "light" scent trails for a moment before returning to a nexus location of scent, and following a second or third scent pathway, until the most distinct trail, or the trail of predominant interest, is definitively identified. Unlike a human, however, the dog can never be potentially misled by interfering ideologies: "Oh, I believe bears are despicable, so this nasty dead deer scent must really be the mark of a bear". The indelible connection between the array of specific deer scents, the indexical representations, and individual deer in particular places at certain times (the representamen), create the highly successful use conditions for a Coon Hound's super scent-grounded sensitivities.[6]

This is not to say that humans are not differentially enabled in their abilities to map the world in terms of scent. In a fashion that, at times, contradicts her central thesis about the dense and ever-emergent scent-scape fashioned by dogs, Horowitz (2016) argues that humans can develop much more attuned abilities to interpret the world through scent. To some degree, this is true. Certainly, Marshall Islanders are proof-positive of such differentially-focused ways of interpreting the world, though at the cultural level. Spoken Marshallese has dozens of primary scent designata, far exceeding those type-classifications in English, and those categories are deployed in everyday life to lend sophisticated scent-interpretations to individual experiences within this Marshall Islands' world. In one interpretation of the Allied overthrow of the Japanese military forces on Majuro, for example, a man from Laura, Majuro Atoll, notes:

> we could smell the Americans long before they set foot on Majuro (Laura islet, Majuro). Perhaps they were still coming downwind in our direction from Rairek, or perhaps there on the first small islets around the airport (10–15 miles distant), when we knew they had landed on this atoll. In just a short time they would be here. We were so cigarette-starved (*an̄ed*), that we could smell the smoke (of the

Americans' cigarettes), and knew that in only a short time, the Japanese would be gone from Laura (the Allies would overwhelm the small Japanese force stationed there).

In everyday use, scents like the residual smell of fish on a person's body part or eating utensil (*jon̄o*), the smell of feces (*kūn̄*), a fragrant smell (*n̄aj*), or the smell of urine (*kūlōj*), are among the more than 50 distinct scent terms listed by Abo, Bender et al. (1976) in the *Marshallese-English Dictionary*. Even trace scents are designated by a separate term, *kāātet*. While, upon occasion, an Enewetak/Ujelang person may refer to a good or bad smell (*bwiin enno* or *bwiin nana*), far more commonly than for English speakers, scents are described with their scent-specific terms, few of which translate into English in a word-to-word fashion.

Therefore, while individual human scent consciousness can certainly be learned, and while Marshall Islanders position scent far closer to the center of their cultural/environmental consciousness than Americans, undoubtedly dogs' worlds, particularly Coon Hound or Siberian Husky worlds, are several magnitudes and vectoral potentialities more complex than humans due to basic biogenetic capabilities that enable their scent-grounded sensibilities. On the other hand, the scent classification schemas of various Pacific Islanders differ from one another and, also, from cultures in other places and times. Therefore, when attempting to assess the presence or absence of dogs in the Pacific, whether in high island locales, or on atolls, both an ethno-semiotics of scent, and several trans-species semiotics, at the level of breed and of everyday conditions of life, must be taken into account.

Hominidified Dogs

The evolutionary history of dogs points to their extremely intertwined relationship with humans. Mitochondrial DNA evidence suggests that the first clade of dogs began their co-evolutionary waltz with humans around 22.8 thousand years ago with a broader adaptive radiation around 15,000 years ago (Perri et al. 2021). Bensky, Gosling and Sinn, presume that humans began purposefully pursuing dogs as highly valued companions at this latter point in time (they suggest 14,000 BP) (Bensky, Gosling, and Sinn 2013:210). Regardless of those early cross-species interactions, the panoply of distinct designer-dogs witnessed today is the result of "the artificial evolution of dogs" based on human selective breeding over the past 150 years (PBS evolution library 2001). Thus, rather than a rapid period of divergent evolution of dog types resulting from adaptive radiation into distinct environments recently vacated by other species, dog's recent artificial evolution occurred through selective breeding and produced numerous distinct breeds with sharply individual traits and, not infrequently, far too many disease-prone recessive genetic characteristics.

In short, cultural selection often operated in ways antithetical to natural selection, producing designer dogs with traits that might well be biologically maladaptive for an individual dog, though contoured to the needs or fancies of their human companions.

While Enewetak/Ujelang people have not actively contributed to the selective breeding of dogs, they shared with members of other cultures the ethnocentric utilization strategies of canines in their communities. On Ujelang, this meant that dogs were used to alert household groups to the presence of others. They also provided a source of food. Given resurgent waves of starvation that might strike atolls, the presence of dogs in people's lives was quite likely intermittent. While historical records are spotty, dogs do not seem to have had a constant presence on Enewetak or Ujelang until German times (1885–1915). Therefore, on these atolls, people have not had long-standing historical opportunities to learn about or exploit the many olfactory-grounded abilities that dogs provide. These superior scent-based abilities, of course, were among those that proved most advantageous to Arctic-dwelling cultural groups whose lives have been consistently intertwined with dogs for long periods of time.

Most Europeans and Americans in contemporary times find the companionship of dogs, their affective contributions to family life, of central concern. Even though dogs stand as the primary referent for the larger Marshall Islands' category "land animals", *kidu*, their specific characteristics are less well known, or at least less storied, than the ubiquitous pig. And, certainly, manifesting a very densely-saturated social environment of fellow human residents, Enewetak-Ujelang people did not heavily rely on the companionship component of canine relationships to enrich people's social/psychological senses of self. Nevertheless, the shape of human/dog relationships certainly began to change when people returned from Ujelang to the New Enewetak. Soon after that move, the island Council instituted a law that required pigs to be contained at the same time that dogs could roam. Even though dogs were hardly considered integral members of families on the New Enewetak, the shift in their residential conditions vis-à-vis pigs, brought them into closer contact with hominid households at the same time pigs became more interactively detached. Not until Enewetak/Ujelang people began to move to the Big Island in 1991, did more substantial changes in human/canine interactions develop.

At the same time, the astute characteristics of dogs, like those of pigs, were closely observed by Big Island Marshallese. As good ethno-ethologists, Enewetak/Ujelang people have developed reasoned understandings of dogs' highly attuned senses of scent and certainly make note of those abilities whenever they are of social benefit. However, Marshall Islanders expend no cultural effort on the construction of folk classifications that attempt to directly compare the relative acuteness of human versus dog scent-detection abilities. The senselessness of such a comparison derives from the fact that, logically, neither Marshall Islands' dogs nor living humans assert final control over their own actions. Both may be "grabbed" by non-corporeal spirit beings who shape the action outcomes of living dogs and living humans. Therefore, the number of miraculous savior tales that might be attributed by an American to the special intelligences of a Lassie or Rin Tin Tin are not heard among Enewetak/Ujelang people. Rather, their stories are contoured in accord with a distinct Marshallese consciousness, and salvational actions involving canine heroics are attributed to the active intervention of one or another local spirit being that acts through a specific dog (cf. Van der Ryn, this volume, Chapter 10). Thus, while local people recognize the extraordinary scent-based abilities of dogs, their associated navigational abilities on land and, at times, their uncanny ability to detect the unsettled state of emotion-laden relationships among humans, when these inherent characteristics are directed to some desired end, the cause of a specific salvational outcome frequently is attributed to an associated spirit that has governed the dog's actions. Only this type of attribution could, for Enewetak/Ujelang people, answer the question of why other dogs in other circumstances did not also find infants who had crawled into the bush, or rectified other undesirable outcomes each time they occurred.[7] Dogs and humans are not masters of their own Marshallese-constructed bodies. Corporeal beings are not solely controlled from within. All corporeal beings, canid and hominid, are part of a vast ethno-scape of causal force, and only the extraordinary activities of living beings infused with super-animate force instilled by perduring beings from the past can account for why certain actions occurred under certain circumstances at specific moments in the life of the community. In a correlative way, the ontology of all mortal Marshall Islands beings is differently grounded than western scientific understandings of animate beings' demeanor. Dogs do not deviate from other entities—plants and animals, or rocks, clouds, breezes or storms—in this regard.

Big Dogs in Hawai'i

As I have discussed elsewhere (Carucci 2012), three young families from Enewetak, moved to the Big Island in January of 1991. They were searching for new opportunities, not only for education, health care, and employment, but also seeking a location where they could live their lives in the Marshallese style—something impossible on the New Enewetak. By 2016, more than 40 percent of the Enewetak/Ujelang community had moved to the Big Island, a percentage that has increased since that time. Not surprisingly, having re-established human–canine relationships while living on the New Enewetak, and having witnessed similar practices in Majuro, the dog–human relationships that have emerged on the Big Island of Hawai'i largely align with the ways those relationships were contoured on Enewetak. Big Island dogs, however, are often larger, and many people say they keep them to reveal when people (that is, non-Marshallese) are near.[8] In early years, the increased size simply may reflect the cast-off dogs that people found available to them for little cost at the humane society. In more recent times, some younger men, and a few elders, have modeled their dog preferences on local Big Island Hawaiians, who prefer pit bulls, German Shepherds, and other large protective breeds to Miniature Poodles or Chihuahuas. Many of those Hawaiian-owned dogs are staked outside houses to perform their policing functions and lack much human interaction. Similarly, for Ujelang/Enewetak folks, most dogs are not well cared-for, much less indulged. They do, however, freely patrol household land parcels. As on Enewetak, at feasting events, dogs are frequently chased off with hissing sounds (signaling "vamoose", [or else]). Should they not leave, persistent dogs are greeted with volleys of rocks. Sometimes, chunks of lava are tossed lackadaisically toward a canine by community members attending the communal food event. At other times, however, stones are thrown with force, eliciting yelps and, occasionally, leaving an animal with life-long injury. When disciplining dogs with volleys of stones, young men in particular, often shout *e-lel*, "it has been struck"—the same incantation uttered when successfully skewering fish or birds on fishing and hunting ventures. Unless the stoning results in serious injury, community members respond to such dog disciplining without sympathy or, particularly when young men are the instigators, with schadenfreude.

Other Big Island dogs, however, have quite different lives, reflecting a different consciousness, one in which the dog, occupying the class *nejiō*, "birth-class endeared being/thing" becomes a social being with whom one interacts with compassion and caring. Best known among these new age canines was Joniten's dog, Bo, who lived a long, relatively-indulged life in Ocean View.[9] Bo finally died in 2019, at least 11 years of age. Bo certainly is not the only dog living a life of increased luxury on the Big Island. Neiwan and Bilimon, respected elders and long-standing Enewetak/Ujelang Big Island residents, had a large black northern-breed-cross named Lava who was also fed and treated humanely. Unfortunately, Lava ran across the Mamalahoa Highway adjacent to the house and met his demise at about 3 years of age. While other dogs were dismissed and abused by Big Island Marshall Islanders, Bo, Lava, and a few additional "*kidu in pālle*" ([indulged/pampered] dogs belonging to Americans/white folks), as other Marshallese referred to them, are kept by Big Island Marshall Islanders.

In spite of the appellation, Joniten strictly disagreed with people's "*kidu in pālle*" claims, recognizing that his "pampering" paled in relation to the highly-indulgent treatment of pets by some Americans he met. He notes that Bo just ate *mōttan mōn̄ā in armij*, the scraps that remain when humans have eaten, while haole (non-native, white) residents on the Big Island feed their dogs separate food. He also noted that Bo slept in the cookhouse, not in the house. When I said, "And not in your bed?" he just scoffed and laughed, "Yeah, you're right that *di pālle* are very crazy." At the same time, Joniten and Tarike were quick to adopt practices they viewed as beneficial. After Bo's death, Joniten began feeding his replacement dogs kibble. When I asked about his comment regarding Bo and "*di pālle*" dog food, he said, "Costco. Less cost (for) this food, the food-class object of dogs, than (for) food-class objects of humans."

Even though Joniten separated his interactions with Bo from the ways many haole residents indulged their dogs, he was proud of the fact that Bo obeyed commands. When Bo came too close during a public food event, Joniten simply pointed and said "Bo, out (in English)" and Bo would leave the eating area. Commenting on Bo's demeanor, Joniten said he was very smart. "There are some kinds of thoughts in his mind. It is not like a human; none-the-less, he is smart." And after his death, Joniten said that, as Bo became very decrepit, Joniten had to take him out so he could pee and poop (indicating he carried him out). The day before he died, however, Joniten said that Bo went out the back of the cookhouse on his own, even though he struggled. He did not come home. When he did not return by mid-day on the following day, Joniten went to look for him and sent some young men to look as well. Finally, the young men found him on the lot south of Joniten's and Tarike's cookhouse. Joniten said:

> Well, as I told you, that dog was smart. Perhaps he knew, and yet he did not want me to know, that he was going to go and lie down and die. Perhaps he was thinking, 'well, it will never be good with Joniten if I make uncertainties for him and for her' (Tarike). So, he struggled, and tried hard, and went out the back and searched for a place where he could lie down and rest.

In this instance, by rest, *kakije,* Joniten means not only a place for Bo to relax, but also "final resting place". This latter meaning may well not be of ancient origin, but it has been in common use at least since the 19th century translation of a missionary hymn that discusses dying as "*buk an kakije*" (to take one's rest).

For Joniten, Bo's intelligence separated him from mistreated dogs in the Big Island Marshallese community. He says, "Well, he was a dog, but really smart. And the other dogs, they are slightly smart, but also crazy (or dumb) like a pig." Of course, in other contexts, Ujelang/Enewetak people say pigs are really smart. None-the-less, they use a pejorative pig classification when chastising young children to say, "Well, perhaps you think you are smart, but you are crazier than a pig." Therefore, not entirely unlike Americans,[10] the human-like characteristics of dogs, or certain canines, a component of their sentient being, makes those canines worthy of special treatment. For Joniten, Bo was the quintessential example of such hominine characteristics.

As a lab-cross, Bo was highly bonded to Joniten, something that elicited the "kidu in pālle" (dog of white people) appellation by other Marshallese community members. When Joniten moved from one location to another, Bo followed him to the new location and laid down nearby. Joniten knew this differentiated Bo from most other community members' dogs. Indeed, Tarike, Joniten's wife, attributed their close relationship to reciprocity: "Well, Joniten really loves that dog, but he (Bo) also watches Joniten, and he protects this house from strangers. He reveals if anyone is approaching the house" (cf., Dobrin, this volume, Chapter 2, for comparable stories from New Guinea residents about dogs as valued exchanging partners). But Joniten did not attribute their close relationship to special behaviors of Bo. Rather, he said:

> Well, the reason that that dog (there by you) really loves me is because I do not abuse him (*mani*). I don't throw (rocks at him) repeatedly and, while I do not give him food-class things like Americans do, I give him left-overs whenever there is a lot of food.

From Food to Friend

Being relatively "new" creatures on the landscape on account of their extirpation, dogs, for Enewetak/Ujelang people, are still positioned as both prototypical and yet distinct in specific ways from other land animals living with and/or around them. But, in people's recollection of renowned canines, Bo is certainly one of the most distinctive experimental dogs to draw the attention of community members who encountered him. He is the prototype for the dog-as-friend potential of the species. In certain ways, Enewetak-Ujelang people have been reconstituting dogs as friend, rather than food, for three or four decades. Some local people see this change as a way of separating themselves from being conceptualized as filling the savage slot (Dominy and Carucci 2005) in their interpretation of others' views of themselves (Cooley 1902). Nevertheless, cultural practices are persistent and the contexts of performance and interpretation pervasive. At best, therefore, this transition is constantly contested and incomplete. Nowhere is this more obvious than in comparing Joniten's interactions with and indulgences of Bo, and the community's interpretations of those interactions, with the practices of Americans.

Certainly, on Ujelang, dogs were never fed anything more than the occasional scraps of discarded human food; largely, they had to forage for themselves. In contrast, Joniten's recent reliance on dogfood to feed Bo's canine successors, represents a different form of consciousness. That shift in consciousness requires Joniten to consider the dog's needs and, perhaps, even the dog's point of view. One afternoon in January 2020, Joniten was retrieving a tool from his tool shed, where he also stored the stash of dog food. As he spread some dog food across the rocks in his driveway, he said:

> It is not good if they are hungry. They will never be peaceful unless you give them some food (literally, "a food-class thing"). They are not dogs from Ujelang. I want them to stay here to watch over this land parcel and see that no one damages the infants and children.

Therefore, in the Big Island context, with neighbors of all stripes nearby, Joniten has begun to contrast his dog feeding with practices on Ujelang and Enewetak where canines, at best, received an occasional left over. Like most Marshall Islanders on the Big Island, Joniten has become more attuned to outsiders who might harm Marshallese. Therefore, in Ocean View, Joniten highlights his dogs' first alert capacities. Without food, he fears the dogs might well become feral, something he witnessed on Ujelang. Instead of risking that outcome, Joniten clearly wants his new dogs to mimic Bo, and remain meaningful parts of his household.

At the same time, Joniten sees dog food as a relatively inexpensive provisioning option for these household members. When I commented on the large bags of dog food, asking if the dog chow was not costly, Joniten said, "Yes, but much less expensive than (providing them) food-class cow and chicken." Even this cost balancing, however, entails a shift in consciousness. It requires that Joniten think of the dogs' needs at the time he purchases provisions for other members of his family. While perhaps not requiring everyday meals of chicken or pork, as contributing members of the household, the family's dogs deserve a share of food. And certainly, for Marshall Islanders, receiving food every day, itself marks one's insider status. In this case, it delineates the dogs' membership in the household.

Certainly, many Big Island community members thought it somewhat humorous that Bo and Joniten were closely bonded. At the same time, other Marshall Islanders were both intrigued by their relationship and envied it. This is readily seen in common statements like *Iokwe, leo Joniten im kidu ne* "Love, welcome, saturated sentiment (goes to) Joniten and that dog there (near you)", followed by "they are very close", or "(their relationship) is of a certain kind (special/notable)". No dog/human relationships encountered on Ujelang or Enewetak were ever discussed in these terms.

Even though being part of a household, as was Bo, marks a critical shift in consciousness, it is not as if Marshall Islander's dogs are taken to the veterinarian for vaccinations or other care. Not infrequently, they limp around after being mauled in an encounter with a neighbor's canines. Moreover, when Tarike asked me about the miniature beds found in a second-hand store in Kona, I told her they were pet beds, "for dogs and cats". Tarike and her fellow shoppers laughed uproariously. "Is it true that these things are for animals/dogs (*kidu*)?", she replied. Their laughter only increased when I confirmed that her query was true, marking the absurdity of the entire idea. And, of course, the thought of an industry built on pet toys, to say nothing of pet hotels where owners of dogs and cats might pay hundreds of dollars to house their pets for a single night's stay, verges on the unimaginable. Therefore, if dogs like Bo are inching toward a new positionality, something similar to "pets", even for Joniten and Tarike, the limitations on that symbolic re-categorization remain quite substantial.

On the other hand, Bo was buried on the edge of Joniten's and Tarike's land, and a small cairn marks the location. In no way does this replicate the costly pet headstones in a variety of pet cemeteries in the United States, but it does entail a form of recognition and care unlike the dogs on Ujelang, whose bones were strewn among those of pigs and chickens, post consumption, on the ocean-side of the islet. In ancient times on Ujelang and Enewetak, people recall that deceased humans were floated out to sea in small outrigger canoes in an attempt to distance their non-corporeal spirit selves from inhabited sections of an atoll (Carucci 2016). Beginning in missionary times, however, family graveyards came to be located on a lagoon-side or ocean-side margin of an extended family's land parcel. While such family memorial practices were prohibited when people moved to Kona, Hawai'i (Carucci 2016), replaced by burials in the nearby Na'alehu cemetery, Bo's marked burial location designates a place of distinction along the margin of Joniten's and Tarike's land. Ironically, it replicates the desired practice that community members would like to follow on the Big Island for the internment of their fellow human consociates.

Bo's Perspective

While it is impossible to know exactly how Bo believed he fit into Joniten's and Tarike's household, undoubtedly, he was aware of his special position. He entered their large cookhouse with confidence, quite in contrast to other dogs, who would cower and lower their ears as they approached any human dwelling. Of course, the common response for dogs was to be greeted with a sharp human utterance "āāh! Sssh!", scaring them away from the scene. And any dog that challenged that response would be chased off with a volley of paving pebbles or other objects hurled in its direction as it turned to race from the scene. During very special human events, particularly funerary meals, Bo might be hissed out of the cookhouse, and he would readily depart with a slight hissing and flick of the wrist, as Joniten motioned him out. But Bo knew these times were exceptional. He lowered his ears when Joniten gestured him out, but Bo did not cower or tuck his tail as he departed. He knew no volley of stones would follow as long as he followed Joniten's command. Even for large *keemem,* first birthday's celebrations, that were frequently held in Tarike's and Joniten's oversized cookhouse,[11] Bo was allowed to attend as long as he lay calmly near Joniten's feet. For Marshall Islanders not familiar with core Enewetak/Ujelang standards, Bo's presence garnered some special side-commentary, but core community members did not question his presence. They knew that Bo was a meaningful part of the household.

Bo's renown contributed to his interaction style outside of the cookhouse and rainy-day work area. As with most Marshallese dogs, he barked when others parked and began to walk up the driveway. Community members who knew Bo best hardly paid him any mind, simply glancing in his direction. Others might warn him away: "Ah, Bo" and hiss slightly. Bo would simply stop barking and move away. Or, if Bo did not recognize the visitor, he would retreat to the cookhouse and jump inside. Then, facing out to watch the approaching humans, he would growl softly until Joniten looked up and silenced him *Ah, Bo, nana* "Ah, Bo, don't do that", or *Ah, Bo, ta ne?* "Ah, Bo, what is it (there by you)?" Bo was more confused when I approached the house since I would speak to him directly, often before Joniten silenced him, and would scratch his ears if he came close. Not surprisingly, Bo learned that he could stop by and be petted if he came close to me, even if we were eating food. While others might give him a bone or other discarded scrap during a food event, Bo frequently would come first to me, knowing he would likely receive food and a pat on the head. While children occasionally tested out my "patting/petting" theory, no other adults ever petted Bo or scratched his ears. While a living creature, he was considered a different-order being, potentially polluting and, therefore, hesitantly welcomed. As much as I was considered a fellow community member by many, scratching Bo's ears definitely marked me as a different type of human, one with residual *di pālle* (clothed-person's) demeanors. At public events, other adult men or women would just smile slightly if they saw me scratch Bo's ears. Still, from Bo's perspective, approaching me was always less inhibited than approaching others for food. He had no ambivalence, since he knew he would receive some positive response, a pat or food. With others, both men and women, he might be chased away or receive a scrap of food.

None of Joniten's and Tarike's replacement dogs have risen to Bo's level of acceptance and, to date, they are most commonly chased out of the cookhouse when they attempt to enter. But they hover around the door when food events are in progress realizing that they will not be chased off unless they transgress the boundary marked by the cookhouse entryway. Therefore, from the dog's perspective, I believe we learn that pethood is hardly a rigidly-defined category, either for Marshall Islanders experimenting with an innovative interspecies relationship, or for the members of American households. Levels of welcome and indulgence vary, but that does not prevent Americans from saying that they own a pet, even if the dog is not permitted to occupy the main bed in the master bedroom, or place his/her chin on the table at suppertime. While functionally distinct, perhaps Bo's special status in Joniten's and Tarike's household is most like a farm or ranch working dog, but one amply indulged to be allowed in the house during eating and sleeping events, even if not given a chair at the table or given permission to occupy the mistress' (or master's) bed. Bo was a special household member with distinct privileges and indulgences, however restricted in contour. His positionality was secure and recognized well beyond the household. The site of Bo's passing remains marked, though without a plaque or name attached to the locale.

In all of my questions about family kinship links, even Bo was never given a slot in Joniten's and Tarike's family, in contrast to the dogs or cats that appeared in the kinship charts of some members of my Introductory Anthropology classes. Not infrequently, those students—given the Schneiderian guidelines with which they pursued their assignments (focused on meaning and significance, not "blood")—listed their pets as significant family members. Therefore, if not "pet" (for which there is no specific Marshallese category), Bo was a significant household member, storied even after his passing. He was designated by others as *nejin Joniten* "that offspring-class personage of Joniten's" and earned his keep by protecting members of the household and showing them affection. This was especially true of Bo's actions toward Joniten. Their relationship was bound by shared sentimentality across interspecies boundaries.

Joniten's recognition that his gentle treatment of Bo resulted in a high level of trust on Bo's part is undoubtedly true. However, highly domesticated dog breeds are not only dependent on humans, they show evidence of attachment behaviors to humans that are quite pronounced when compared to the "social affiliative" behaviors of wolves, even young wolves raised to enhance their level of comfort with humans (Gácsi, Miklósi, and Topál 2023). Certainly, wolves can perform activities that require intelligence and cunning, much like their dog offspring/cousins, but there is a substantial body of evidence that their attachment behaviors, a phenotypic trait that Gásci, et al. contend, developed after the human-dog co-domestication began, is quite distinct from dogs. The Strange Situation Test is used to measure dogs and wolves sense of comfort around humans, and Gacsi, Miklósi, and Topál (2023) note how critical it is to administer this test with utmost caution, since the results of one recent study (Hansen Wheat et al. 2022) suggest that young wolf pups, raised by humans manifested these same traits. Hansen Wheat, et al., contend that their research not only suggest, but "demonstrate that the ability to form attachment with humans … refutes "claims that this phenotype evolved after dog domestication was initiated" (2022: e9299; abstract and p. 10). Gacsi and her research colleagues suggest that the Hansen Wheat study is inadequate to overturn the evidence from multiple earlier studies that have relied on the Strange Situation Test, noting that one problematic parameter of the study was its reliance on two litters of puppies from a very special breed, Alaskan Huskies (Gacsi, Miklósi, and Topál 2023). Further research undoubtedly will help clarify the significance of the Hansen Wheat et al. (2022) study. Indeed, in a recent Science article, Hansen Wheat notes that she and her fellow researchers are "not arguing that dogs and wolves are the same … .What we saw does not make (these young wolf pups) dogs" (Grimm 2022). One parameter requiring further study will be to examine the difference between huskies and golden retrievers, or similar highly domesticated breeds, inasmuch as malamutes, huskies, akitas, etc. still share behavioral residues of their wolf ancestry. This is hardly surprising since, as Perri et al. (2021:3) estimate, based on genetic evidence, the A1b clade of dog ancestors separated from the A2 clade somewhere around 22.8 ka, during the last glacial maximum, and well before the A2b radiation of dog types that had migrated into North America at around 15 ka. Many current northern breeds share characteristics unique to this A1b clade. Hence, even though huskies, like many other northern breeds, are perfectly happy to share the same spaces as their human housemates, they are renowned for having minds of their own: they can be trained to do the same tasks as other dogs, but are not absolutely obedient. When determined to follow their own path, no amount of training will dissuade these early-clade wolf offspring from doing what they wish. Moreover, while sharing spaces of human design, and while capable of showing great affection, they are also happy to be distant observers of human activities and, only by their choice, cuddle closely with their humans. Like wolves, they are intelligent and cunning, but not blindly enamored with their nearest human companions. Bilimon's and Neiwan's dog was a husky-mix. Therefore, it is not surprising that he, like other dogs from so-called "primitive dog" breeds, did not display the same attachment to members of his human family as did Bo (see Sumridge et al. 2021 for similar findings).

When Lava inhabited Bilimon's and Neiwan's home, these "near, but not overly dependent" characteristics typified his demeanor. More than once, Bilimon inquired why Lava was not like Bo. Rather than attributing Lava's relative aloofness to the characteristics of his breed, Bilimon thought that, perhaps, he was just dumb:

> Well, why is it that that dog over there (distant from speaker and interpretant) is not like Bo, and (why doesn't he) lie down here next to me? All the time, he lies around there (at a distance) and if there are food scraps, we throw them out and that dog, Lava, will stand up and go and eat them, and then go and lie down again over there at a distance. Perhaps his mind is a little damaged. Perhaps he drank too much when he was small, at the time he was (a) "puppy" (in English) for a while, and that is why his brain is damaged! (Bilimon joking.) The thing is, he is greatly different from Bo. He will never come and lie down and press my feet into the ground when I am resting and sitting for a while. But all the time, if Joniten is there (next to you), Bo will be lying next to him.

Thus, while Bilimon's observations were astute and absolutely correct, his causal interpretations, like most interpretations of dog demeanor, are anthropocentric and culturally-biased. In the words of Panagiotis Xouplidis:

> the anthropomorphic code (for interactions with dogs) is closely related to the experiential code of pet owners practicing pet-keeping in the modern world and to the social support code from the same individuals as a result of the interaction with their pets
> (2014:1.3.1).

While an inter-relational, interspecies, approach aims to capture the perspective of dogs themselves, leading toward the olfactory-centric interpretations of Horowitz and others mentioned above, Xouplidis believes that humans cannot escape the fact that all semiotic interpretations, including ideas about species themselves, are re-contoured into anthropomorphic codes and languages (2014:1.2 and 1.3.1). In Kull's terms, no matter how much humans follow the lead of Jakob von Uexküll's attempts to capture the "*umwelt*—the study of subjective worlds (or meaningful behavior) of organisms of many different species," all communicative categories and identities contain classic (bio-)semiotic contradictions, thereby requiring, or enhanced by, complementary descriptions (Kull 2008:3211 & 3213). That is, semiotic "interpretations of" can only work to describe in imperfect terms the bio-interactions of trees in their complex environments (Simard 2021) or the *Umwelt* of dogs in their multi-layered interactions. My own positive feelings about dogs only add further layers of semiotic complication.

Bilimon's anthropocentric attributions of cause to Lava's actions are hardly limited to Enewetak/Ujelang Marshall Islanders. Even in their extensive academic overview of dog cognition research, Bensky, Gosling, and Sinn (2013), trained cognitive scientists, contend that they are interested in the ways that "dogs perceive and organize the world around them". Nevertheless, their extensive review of the literature selects in favor of studies that stress visual cognitive tasks rather than olfactory-based activities (2013:303). Therefore, in spite of their own claims, Bensky, Gosling, and Sinn hardly capture dogs' *Umwelt*. If finely-tuned semiotic modeling of dog perceptions and conceptual organizations is possible, it clearly presents extraordinary challenges to even the most attentive and skilled of human interpretants.

Bilimon and other Enewetak/Ujelang people do not have the historically-deep experiential interactions with dogs that is typical of the centuries-long observational opportunities of Inuit and other Arctic dwellers. Neither do most Marshall Islanders have the long-term academic training of some canine researchers, nor do Big Island Marshall Islanders, who frequently work long hours to make ends meet, have unrestricted time to interact intimately on an everyday basis with their dogs. Never-the-less, Big Island Marshall Islanders now discuss their dogs, their everyday activities, and what they might be thinking, far more than did their predecessors on Ujelang or Enewetak in the past half century. Their increasing experience with dogs certainly aligns with dogs' discursive centrality in the affairs of everyday life on the Big Island. Given peoples' understanding of the immense value that an extensive knowledge of fish and bird demeanor provided for them when they lived in intimate relationships with the sea, the land, and the movements of air, clouds, and stars in the sky, increasing levels of interest and understanding between Enewetak/ Ujelang dogs and their humans will undoubtedly emerge in the future.

Notes

1 In comparison to high islands, resources on atolls are always limited. Times of famine, and certainly the ecocide resulting from U. S. nuclear testing on Enewetak and Bikini, increase this dilemma. As Grieg notes (this volume, Chapter 5) one Polynesian archaeological study by Bay-Peterson suggests that there is a "complex balance between human nutritional requirements and the production of surplus from which to feed animals". Not only was this a problem on the "New Enewetak", undoubtedly it was a factor during times of famine on Ujelang, when people ate dogs rather than allowing canines to share foods on which they depended for survival. This perspective aligns with Cramb's long-term assessment of dogs throughout the Marshall Islands. While dogs existed in the Marshalls in pre-European times, they were 'extripated' prior to European contact. Cramb suggests that this pattern is quite likely in atoll settings where the availability of resources is at issue, where dogs compete with humans for those resources based on limited land size, modest species diversity, and the risk of devastation from typhoons and other natural apocalyptic events, which is more common than on high islands (Cramb 2021:33).

2 In contrast to such minimalism, Bradd Shore writes in *Culture in Mind* (1996) about the multiplicity of cultural models humans use to communicate. While one might argue with Shore's categorizations, his differentiations point to the centrality and elaboration of culture for humans in all of the world's societies: these "models" exist as "public artefacts 'in the world' and as cognitive constructs 'in the mind' of members of a community" (1996:44). Mental models both personal and conventional are complemented by institutionally inscribed collective models. For Shore, human cultural model genres may be linguistic or non-linguistic, (physical) orientational, spatial, temporal, social orientational, diagnostic, expressive/conceptual, or task models, and perspectives vary for observational and performative purposes. Shore's linguistic models, as an example of the density of human cultural activities, are further subdivided: scripts, propositional models, sound symbolic models, lexical models, grammatical models, verbal formulas, and trope models (1996:56). Silverstein would add "ethnometapragmatic" models as well (Silverstein 1993). The point is: human cultural modalities are multilayered, extraordinarily pervasive, and dense when compared to the minimalist approach adopted by Guillo and Claidière as they propose to assess interspecies "culture".

3 For Leslie White, a student of Boas, symbolling (the core human culturally-productive capability) was the most radically transformative capacity encountered in the history of earthly evolutionary occurrences (White 1949).

4 Viveiros de Castro does call this capacity of non-human beings "cultural", but he does so on account of the ability of other animals to place themselves "in the position of subject", and to see themselves as members of (what for dogs) would be a dog-centric "human" species (2016:244–245). But, on these same pages he warns against the "symbolic spiritualization of animals", and this is

precisely the enterprise that Guillo and Claidière, as many others, pursue when they find culture among other animals.

5 While Lohmann (this volume, Chapter 1: 10), strongly believes that dogs are co-participants with humans in inter-species "culture", he agrees that humans are the "cultural species *par excellence*", and that dogs "do not use language among themselves or display intricate and diverse social conventions and symbolic systems akin to human cultures". In contrast, the human use of an arbitrary system of symbolic representation, I argue, is a prototypical definitional feature of culture. Claiming that dog sleds, collars, and leashes are cultural co-productions vastly confuses the whole idea of culture, in my view, as does Lohmann's contention that "interspecies culture … can be recognized in elements of a living thing's genotype that was shaped by a culturally influenced interaction of more than one species" (this volume, Chapter 1: 15). While the author gives domestication as an example of such cultural co-production (since research suggests wolves were active contributors), what about the SARS-CoV-2 virus? Reformulations of the virus' own genotype clearly fit Lohmann's statement, so the virus clearly must be an active co-producer of interspecies culture. By such a definition, each entity in the universe (yet identified or not) participates in interspecies cultural production, rendering any meaningful notion of culture vacuous.

6 Of course, in the presence of their interspecies masters, dogs may learn to replicate certain biases of their owners. Bird dogs, for example, are taught to retrieve and drop birds on command. But, the human teaching strategy involves alternate rewards. Once the human is no longer present, bird dogs will not by themselves pass on to their grand-dog offspring's offspring, any biases about the evils of consuming certain types of ducks or quail because they heard a story from other dogs about the un-groomed character of their plumes, thus making them undesirable. Such ideological biases, without grounding in everyday experiences, are not part of a dog's communicative repertoire.

7 While the causal agents differ, non-corporeal force channeled through ancestral beings for Enewetak/Ujelang folks acts in ways similar to what Evans-Prichard termed "witchcraft" among the Azande. The explanatory power of unique outcomes is similar for both. The logic of why a particular stubbed-toe wound of a boy might fester, why a woodcarver's bowl might crack, or why a granary might collapse on a group of people seeking shelter from the sun, were all caused, for Azande, by witchcraft (Evans-Prichard 1976:21–23). Similarly, why a certain dog at a specific *kemem* might lead young children directly to a wayward toddler when other dogs in different circumstances did not, could not, for Enewetak/Ujelang people, be adequately explained by the superior scent-detection capabilities of the dog. Those capabilities were necessary, but not sufficient in-and-of themselves, to lead to the child's discovery.

8 A dog's "early notification" abilities on the Big Island are similar to people's stated valuation of dogs on Enewetak. However, the fact that Big Island visitors might well be strangers altered the value of Big Island dogs. Even though the transition was not immediate, dogs' early notification also contributed to Marshall Islanders seeking out larger, and potentially more fear-inspiring, dogs.

9 Joniten and his wife, Tarike, are one of the three founding settler families of the Big Island Enewetak/Ujelang community, first arriving in January 1991. As one of the leaders of the Big Island community and an astute reader of how Enewetak/Ujelang people have been received by others, I believe Joniten was eager to demonstrate that Marshallese, too, could treat their animals with respect. This aligns with Joniten's attempt to ensure that outsiders do not falsely presume that Marshall Islanders are "primitives", but are at the same "level" as *di pālle* "clothed people", or "white folks".

10 Now, nearly 35 years after Joniten and Tarike arrived on the Big Island, many members of the Marshall Islands' community **are** native-born Americans, citizens of the United States. Older generation Marshall Islanders living in the United States are legal foreign residents under I95 passport regulations. Nevertheless, due to the othering of Marshall Islanders by Big Island residents, and to the welcoming embrace and inclusionary practices by members of the Big Island Enewetak/Ujelang community, local Big Island Marshall Islanders consider themselves diMajel, never diPalle (clothed [white] people, or Americans).

11 Their original cookhouse was transformed into the community's UCC church locale. Their cookhouse then moved to a large specially-constructed dwelling nearby, originally designed for Sunday School classes, and subsequently expanded to include a cooking section and large seating area for community events like *keemem*.

References

Abo, Takaji, Byron W. Bender, Alfred Capelle, and Tony DeBrum. 1976. *Marshallese–English Dictionary.* Honolulu: The University Press of Hawaii.

Bensky, Miles K., Samuel D. Gosling, and David L. Sinn. 2013. The world from a dog's point of view: A review and synthesis of dog cognition research. *Advances in The Study of Behavior* 45: 209–406.

Carucci, Laurence M. 1980. The renewal of life: A ritual encounter in the Marshall Islands. PhD. Dissertation. Department of Anthropology, The University of Chicago. Chicago, Illinois.

Carucci, Laurence M. 1987. Jekero: Symbolizing the transition to manhood in the Marshall Islands. *Micronesica* 20: 1–17.

Carucci, Laurence M. 2011. From famine to feast: The give and take of food in a global subsistence environment. *Pacific Studies* 34(1): 1–23.

Carucci, Laurence M. 2012. You'll always be family: Formulating Marshallese Identities in Kona, Hawai'i. *Pacific Studies* 35(1–2): 203–231.

Carucci, Laurence M. 2016. Fearing the dead: The mortuary rites of Marshall Islanders' amid the tragedy of Pacific modernity. In *Mortuary Dialogues: Death Ritual and the Reproduction of Moral Community in Pacific Modernities.* David Lipset and Eric K. Silverman, eds. Pp. 25–46. Oxford: Berghahn.

Clifford, James. 1983. On ethnographic authority. *Representations* 1: 118–146.

Cooley, Charles H. 1902. *Human nature and the social order.* New York: Scribner.

Coren, Stanley. 1994. *The intelligence of dogs: Canine consciousness and capabilities.* New York: The Free Press.

Cramb, Justin. 2021. The dogs of Remote Oceania: An archaeological and ethnohistorical view of domestic dog introduction and loss in the South Pacific. *Archaeology in Oceania* 57(1): 28–38.

Dominy, Michele, and Laurence M. Carucci. 2005. Anthropology in the savage slot: Reflections on the epistemology of knowledge. *Anthropological Forum* 15 (3): 223–234.

Etymological Dictionary online at https://www.etymonline.com/word/humane.

Evans-Prichard, E. E. 1976. *Witchcraft, Oracles, and Magic among the Azande.* Oxford: Oxford University Press.

Gácsi, Marta, Ádám Miklósi, and József Topál. 2023. Comment on "Human-directed attachment behaviour in wolves suggests standing ancestral variation for human-dog attachment bonds". *Ecology and Evolution.* doi:10.1002/ece3.10514.

Geertz, Clifford. 1973. *The invention of culture.* New York: Basic Books.

Grimm, David. 2022. Can wolves bond with people like dogs do? *Science.* doi:10.1126/science.ade9589.

Guillo, Dominique, and N. Claidière. 2020. Do guide dogs have culture? The case of indirect social learning. *Humanities and Social Sciences Communications* 7 (31): 1–9.

Haines, Helen R., and Clare A. Sammels, eds. 2010. *Adventures in eating: Anthropological experiences in dining from around the world.* Boulder: University Press of Colorado.

Hansen Wheat, Christina, Linn Larsson, Patricia Berner, and Hans Temrin. 2022. Human-directed attachment behavior in wolves suggests standing ancestral variation for human-dog attachment bonds. *Ecology and Evolution* 12(9):e9299.

Haraway, Donna. 2003. *The companion species manifesto: Dogs, people and significant otherness.* Chicago: Prickly Paradigm Press.

Haraway, Donna. 2008. *When species meet.* Minneapolis: University of Minnesota Press.

Hare, Brian, Michelle Brown, Christina Williamson, and Michael Tomasello. 2002. The domestication of social cognition in dogs. *Science* 298: 1634–1636.

Henry, Jules. 1963. *Culture against man.* New York: Random House.

Hill, Erica. 2018. The archaeology of human–dog relations in Northwest Alaska. In *Dogs in the north: Stories of cooperation and co-domestication.* Robert P. Wishart, Jan Peter Laurens Loovers, and Robert J. Losey, eds. Boca Raton, FL: Routledge.

Horowitz, Alexandra. 2016. *Being a dog: Following the dog into a world of smell.* New York: Scribner.

Kohn, Eduardo. 2013. *How forests think: Toward an anthropology beyond the human.* Berkeley: University of California Press.

Kull, K. 2008. Semiotic ecology. In *Encyclopedia of Ecology.* Sven Erik Jørgensen and Brian D. Fath, eds. Pp. 3210–3214. Amsterdam: Elsevier.

Lévesque, Francis. 2018. Threatening the fantasy of an Arctic welfare state. In *Dogs in the north: Stories of cooperation and co-domestication.* Robert P. Wishart, Jan Peter Laurens Loovers, and Robert J. Losey, eds. Boca Raton, FL: Routledge.

McCormack, Patricia A. 2018. An ethnohistory of dogs in the Mackenzie Basin (western Subarctic). In *Dogs in the north: Stories of cooperation and co-domestication.* Robert P. Wishart, Jan Peter Laurens Loovers, and Robert J. Losey, eds. Boca Raton, FL: Routledge.

Musharbash, Yasmine. 2017. Telling Walpiri dog stories. *Anthropological Forum* 27 (2): 95–113.

PBS evolution library. 2001. Evolution of the Dog. Available online at: pbs.org/evolutionlibrary.

Peirce, Charles S. 1960 [1931]. *Collected papers of Charles Sanders Peirce* (2 volumes). Cambridge, MA: Harvard University Press.

Perri, Angela R., Tatiana R. Feuerborn, Laurent A. F. Frantz, Greger Larson, Ripan S. Malhi, David J. Meltzer, and Kelsey E. Witt. 2021. Dog domestication and the dual dispersal of people and dogs into the Americas. *PNAS* 118(6): 1–8.

Povinelli, Elizabeth A. 1995. Do rocks listen? The cultural politics of apprehending Australian Aboriginal labor. *American Anthropologist* 97(3): 505–518.

Poyer, Lin, Suzanne Falgout, and Laurence M. Carucci. 2000. *The typhoon of war.* Honolulu: University of Hawai'i Press.

Sahlins, Marshall. 1976. *Culture and practical reason*. Chicago: The University of Chicago Press.

Savishinsky, Joel. 1975. The dog and the hare: Canine culture in an Athapaskan band. In *Proceedings of the Northern Athapaskan Conference, 1971: Volume 2*. Annette McFadyen Clark, ed. Pp. 462–515. Ottawa: National Museums of Canada, University of Ottawa Press.

Shipman, Pat. 2015. *The invaders: How humans and their dogs drove Neanderthals to extinction*. Cambridge, MA: The Belknap Press.

Shore, Bradd. 1996. *Culture in mind*. Oxford: Oxford University Press.

Silverstein, Michael. 1992. The indeterminacy of contextualization: When is enough enough? In *The contextualization of language*. Peter Auer and Aldo Di Luzio, eds. Pp. 55–76. Amsterdam: John Benjamins.

Silverstein, Michael. 1993. Metapragmatic discourse and metapragmatic function. In *Reflexive language: Reported speech and metapragmatics*. John A. Lucy, ed. Pp. 33–58. Cambridge: Cambridge University Press.

Simard, Suzanne. 2021. *Finding the mother tree: Discovering the wisdom of the forest*. New York: Alfred A. Knopf.

Stocking, George W., Jr. 1968. *Race, culture, and evolution*. New York: The Free Press.

Sumridge, Molly H., Malini Suchak, and Christy L. Hoffman. 2021. Owner-reported attachment and behavior characteristics of New Guinea Singing Dogs living as companion animals. *Anthrozoös* 34(3): 375–392.

Viveiros de Castro, Eduardo. 1998. Cosmological deixis and Amerindian perspectivism. *Journal of the Royal Anthropological Institute* 4: 469–488.

Viveiros de Castro, Eduardo. 2004. Perspectival anthropology and the method of controlled equivocation. *Tipiti: Journal of the Society for the Lowland South America* 2: 3–22.

Viveiros de Castro, Eduardo. 2016. *The relative native: Essays on Indigenous conceptual worlds*. Chicago: HAU Press.

Weisler, Marshall I. 2001. Precarious landscapes: Prehistoric settlement of the Marshall Islands. *Antiquity* 75: 31–32.

Wishart, Robert. 2018. Telling stories of co-domestication and cooperation: An Introduction. In *Dogs in the north: Stories of cooperation and co-domestication*. Robert P. Wishart, Jan Peter Laurens Loovers, and Robert J. Losey, eds. Boca Raton, FL: Routledge.

White, Leslie. 1949. *The science of culture*. New York: Farrar, Straus and Giroux.

Xouplidis, Panagiotis. 2014. A semiotic approach to the pet world. *Proceedings of the World Congress of the International Association for Semiotic Studies*, Sofia, Bulgaria. IASS Publications & NBU Publishing House. https://iass-ais.org/proceedings2014/view_lesson.php?id=18.

10 A Samoan Odyssey

Mysterious Spirit Dogs and Interspecies Enculturation

Micah Van der Ryn

Introduction

My initial motivation to join this volume's collective effort to describe dogs and their humans in Oceania began with a desire to solve an unsolved mystery that harks back to 1980 when I was 19 years old on a solo circumnavigation walk around Savai'i, Samoa's largest island. One moonless, pre-dawn morning, unbeknownst to me, two Samoan "spirit dogs" mysteriously and silently joined me. An hour later the rays of the rising sun illuminated a large healthy brown dog walking ten feet in front of me, and a smaller white, equally healthy dog walking ten feet behind me. I was baffled. Where had they come from? To whom did they belong? Why did they join me and why would they not leave me no matter how hard I tried to get them to do so? Several hours later a Samoan woman I met on the road told me "*E leai ni maile lenā ae la agaga o mai fia feasoani ia te oe*", roughly translated, "these are not really dogs, but rather, special spirits that have come to help you", what I label in this chapter as "spirit dogs". The two spirit dogs stayed with me for two more days, as my protective guardians, always walking in the same formation—brown dog in front, and white dog to my rear.

Forty years later I am finally trying to solve the mystery of these dogs, and more broadly develop and apply an interspecies enculturation framework to explore how *Canis familiaris* learn to fulfill canine roles in Samoan culture, even on rare occasions becoming "spirit dogs". Some of the underlying questions I explore include: How can the rare occurrence of "spirit dogs" be understood as a special outcome of canine enculturation in Samoa? How common is their occurrence and what conditions give rise to them? Why did these "spirit dogs" choose me and stay with me for two days? What is the ontological basis of the belief in "spirit dogs" in Samoan culture? Why would they not be considered dogs? How widespread is a Samoan belief in "spirit dogs"? What about stray dogs in Samoa; how are they regarded within the society?

I structure my exploration by first establishing key principles and dynamics of a Samoan enculturation framework, then examining how

DOI: 10.4324/9781003658696-10

both humans and canines learn and develop roles in that framework, and finally, I use those understandings to help solve the mystery of the two "spirit dogs". To develop understandings about Samoan enculturation I heuristically employ autoethnography, a qualitative research methodology that can be defined as

> an autobiographical genre of academic writing that draws on and analyzes or interprets the lived experience of the author and connects researcher insights to self-identity, cultural rules and resources, communication practices, traditions, premises, symbols, rules, shared meanings, emotions, values, and larger social, cultural, and political issues
>
> (Poulos 2021:6).

As such, autoethnography involves "artistic and analytic demonstrations of how we come to know, name, and interpret personal and cultural experience" (Adams, Holman Jones, and Ellis, 2015:21–25). I deem the autoethnographic method appropriate in this chapter because I draw on many casual observations from my long-term personal and professional ethnographic experience in Samoa.

My understandings of Samoan culture derive from almost 50 years of personal and professional relationships with Samoan people, starting from when I was 11 living for six weeks with the Taofiloa family in Falefa Village in 1972, to multiple visits during my undergraduate and graduate university years for fieldwork and filmmaking projects, to working as an ethnographic researcher/filmmaker, and teacher American Samoa Community College for 22 years. Then there is also my marriage to Faaiuga Faaofonuu from Sili Village, Savai'i, in 1998 with whom I have had four children, and one grandchild, and then also fulfilling the role of *faiavā* (son-in-law) and then a *matai* (chief) of my wife's large extended family in Sili Village, Savai'i.

Throughout my many years in the Samoan Islands I was able to observe many canine behaviors and roles played in society. However, due to the anthropocentric bias of my anthropology education, I never thought of incorporating those observations into my Samoan cultural understandings or research. Remedying this situation here both widens and deepens my previous Samoan cultural understandings in new and interesting ways. Another caveat is, while I like dogs, I do not consider myself a dog person, and for only brief stints in my life had a pet dog in the family. I thus approached writing this chapter not with an expert mind, but with "beginner's mind", that is, with fresh eyes, and an open mind without preconceived expectations.

Archaeological and Early Historical Evidence of *Canis familiaris* and their Lives in Samoa

Despite a large quantity of ethnohistorical and ethnographic literature about Samoa, the amount of published information about dogs in Samoan culture (both past and present) is scant. None of that scant information is presented as part a systematic interspecies approach. Yet, it is useful information to summarize as a foundation on which to build. Archaeological evidence suggests that first human settlement in the Samoan Islands began as early about 2,900 years ago (Petchey 2001; Kirch and Hunt 1992:230). Those Proto-Polynesians were part of the Lapita cultural complex, which lasted up to 2,000 years B.P. The earliest evidence of *Canis familiaris* in Samoa starts from 1500 to 1400 years ago (late or post Lapita period) and continues to the historic period with European contact in the late 18th, early 19th centuries (Smith 2002:105). This indicates the species' relatively long period in the Samoan Islands, co-evolving with their humans to co-create an inter-species Samoan culture. This evidence does not indicate how dogs were initially introduced to Samoa in the late Lapita period—perhaps either via Samoan voyagers returning to Samoa with dogs after visiting islands to the west that already had dogs, or through voyagers from those islands visiting Samoa. The archaeological evidence also indicates that Polynesian voyagers of Samoa sailing to and settling East Polynesia some 1,000 years ago brought dogs with them (Greig et al. 2018:6).

Early and mid-18th century European explorers to Eastern Polynesia, such as Tahiti and Hawaii, reported the natives relishing dog as part of their diet (Titcomb 1969:34–35). Sustained European contact with Samoan society began later in 1830 with the arrival of the missionary, John Williams. Soon European breeds of dogs were introduced and bred with local Samoan breeds. A latter missionary, Reverend John B. Stair who came to Samoa before 1844 reports dogs "were found on all the islands [of Samoa], but the breeds having become so mixed, it is difficult to say what is the original stock" (Stair 1897:187). With regards to whether dog was part of the Samoan diet, he reports "Dogs were formerly eaten by the Samoans, as at other islands [e.g. Tahiti, and the Hawaiian Islands]; of late years, however, the practice has been discontinued". William B. Churchward, a British consul residing in Samoa in the late 19th century, reported, "No native feels quite furnished until he has one or more curs attached to his establishment" (Churchward 1887:355). In conclusion, it appears that for the vast majority of Samoan history, dogs were not desired as food, but rather for other valuable roles they could fulfill in Samoan households.

In contrast to Samoa, the neighboring Tongans did and do eat dogs, which I first learned about through a joke some American Samoan teenagers told me. They said:

> An American Samoan family went to the airport to pick up a visiting Tongan family. The Samoans took the Tongans to a restaurant and ordered hot dogs for the Tongans. When the hot dogs were served, the

> Tongans asked, "What is this?" to which their American Samoan friends replied, "They're hot dogs." The Tongans responded, "Sorry, we don't eat that part of the dog!"

Curiously, Captain Cook, when he came to Tonga in 1773 reports that he saw no dogs there (Titcomb 1969:26), despite the fact that Samoans and Tongans had frequent voyaging back and forth between their respective islands at the time and before. Only later, Europeans brought Tahitian dogs to Tonga where they reproduced and became mixed with introduced European breeds. However, unlike in Samoa, dogs become a favored part of the Tongan diet.

Perhaps one reason that dogs in Samoa were (and are) not eaten could relate to spiritual beliefs about dogs in both pre-Christian and post-Christian Samoa. During my Samoan fieldwork I both read and heard about various wild animals as incarnated spirits, but never in the form of dogs, until I came across Reverend Peter Turner's (1884:49) report that:

> The dog, some kinds of fish, and birds, were sacred to the greater deities … and of course, all the people rigidly abstained from [eating] these things … Salevao—"sacred one of the bush"—a war god in a number of villages was incarnate in a dog—a white one usually. When he wagged his tail, barked, and rushed ahead in sight of the troops of the enemy, it was a good sign; but to retreat or howl was a bad omen. In some places, Salevao was a general village god as well as a war god, … the name of a family god also, and incarnate in the eel, and the turtle!
> (Turner 1884:49).

Turner's description reflects the idea of a culturally deep–rooted Samoan appreciation of, and trust in, canine's well developed extra sensory social intelligence. Therein lies a Samoan spiritual ontology in which the identity of beings is both permeable and flexible. I see a striking continuity between the pre-Christian Salevao spirit incarnated in a dog in the early 1800s with the "spirit dog" explanation I received from the Samoan woman I met on the road in 1980 in a Christian Samoa, but I could not find any recent Samoan ethnography from the 20th or 21st century reporting on the phenomena of Samoan "spirit dogs".

More Recent Observations of How Dogs Fit into Samoan Society

As household pets, Samoan dogs are named and protected members of Samoan households. In discussing the open sided Samoan *fale* (house), the late 1920s, Māori ethnographer, Te Rangi Hiroa (AKA Sir Peter Buck) observed, "The floor gravel is … useful for expelling pigs and fowls. Dogs had the freedom of the house, but such missiles (stones to weigh down mats) are used against them if their behavior needs reprimanding" (Buck 1930:80). My observations starting 50 years later concur, though I must add that the freedom of dogs in household spaces becomes more circumscribed during formal ceremonial events. But also, I observed that dogs appear to learn when such events are occurring and to keep their proper distance, so as not to endure any physically painful reprimand.

The only serious violence I ever saw a Samoan inflict on a dog was in Falefa. A neighbor's dog had attacked my Samoan family's dog, Rusty, just outside their open sided Samoan house. My Samoan brother, Tana, who was sitting at the house post close to the dog fight picked up one of those large stones for holding down the corner of the mat and threw it squarely at the offending dog, sending it whimpering away. Tana's reaction reflects a fierce Samoan loyalty to protect other members of one's group—including, in this case, the family dog, Rusty. An interesting side point is how the wall-less Samoan house facilitated Tana to immediately identify where the dogfight was occurring, and then, from his sitting position cross-legged on the floor, throw a stone at the offending dog.

Interestingly, when I recently mentioned Rusty, who died some 40 years ago, to Merelini, a member of the Taofiloa family, she reminisced about how nice a dog Rusty was, and how much she misses him. Clearly, this reflects how household dogs are thought of as sentient beings to be protected, and remembered after they die. Yet, they are also low ranking members of hierarchal Samoan households, squarely below in rank than any of the human. Curiously, though I have not researched it, the term "*maile*" (dog) is used in Samoa to insult a person, similarly to in the USA and many other countries of the world.

A Framework for Examining Enculturation of both Humans and Dogs

According to the Merriam Webster Dictionary, enculturation, is "the gradual acquisition of the characteristics and norms of a culture or group by a person, another culture, etc." Another definition the dictionary gives is "the process by which an individual learns the traditional content of a culture and assimilates its practices and values". Using an interspecies approach to view the roles of dogs in Samoan society requires seeing dogs as fitting into the same cultural framework as humans are, but in canine, not human, ways, but on terms primarily dictated by their humans, who are the dominant species in the interspecies culture. We can view how both humans and canines come to fulfill various social and economic roles through identifying underlying principles, dynamics, or premises of that framework. For purpose of my discussion here, I will highlight the dynamic cultural flexibility of roles and identity by which cultural scripts can be fulfilled. I will also discuss the overriding principle of *tautua* (or service) towards collective aims that operate in a hierarchal structure.

Flexibility of Societal Roles in Samoan Life

During her 1920s Samoan fieldwork Margaret Mead noted, "Samoans tend to interpret each situation in terms of a social hierarchy in which their eyes are always on the play, never on the players, while each individual's task is to fit his role" (Mead 1937:286). She illustrated this flexible, relational, context specific principle with the following:

> The skilled carpenter or canoe builder becomes, by virtue of his mastery and control over a given piece of work, a chief for that occasion, and must be addressed with all chiefly honours, although once the occasion is past, he maybe only an untitled youth. Conversely, a high chief who wishes to call upon a *taupou* who is visiting his village may explicitly lay aside his title and resume the title of a young *manaia*. As such, he may behave as he could were he not holding the title of a chief
>
> (Mead 1937:286).

Anthropologist Bradd Shore, who did his primary fieldwork on Savai'i in the 1970s also noted this relational, flexible, context specific dynamic, stating, "Samoans focus on the contextual grounding of experience as opposed to things unto themselves" (1982:136). The question now arises if this enculturation dynamic is one that Samoan dogs also come to learn and adopt.

Following the autoethnographic thread of my exploration I wish now to discuss a few examples of how I learned how this highly relational, context specific cultural dynamic, which together with the Samoan need to articulate social hierarchy in many social situations, runs counter to my primary enculturation in America where individual autonomy is emphasized, such that things (people, animals, objects, etc.) gain identity into and unto themselves, and where hierarchy tends to be downplayed in cultural scripts.

Example 1: (From my six week stay with the Taofiloa family in Falefa in 1972 when I was 11 years old.) Mealtimes are an important occasion for expressing social hierarchies. Chiefs (*matai*), elders, and honored guests, designated as higher ranking, sit on the front side of the house and eat first, served by younger, lower ranking, untitled people facing them from the house's rear side. Trays and dishes are loaded with the various foods, and servers keep loading more as food is eaten, to give the impression of unlimited food to be consumed. When those eating declare they are full, servers take away the trays and dishes of half eaten food, replacing them with bowls of water and hand towels for the higher ranking to wash and dry their hands. Only after all those higher ranking have completed eating and washing their hands, do the serving group eat. Soon the household dogs gather outside on the house's rear, knowing their turn will be next. One of the servers will take the trays outside to give the dogs left over scraps. Thus, everyone, including the household dogs, knows their place in the temporal and spatial order. Here, the open wall-less Samoan architecture also plays an important role, particularly when people sit on the mats on the floor, not at tables and in chairs (a subject elaborated upon in my doctoral research; see Van der Ryn 2012).

During the six weeks that my classmate Rufus and I lived with the Taofiloa family, we were consistently treated as honored guests at mealtimes. Hence, Rufus and I were always told to sit on the mat at the front side of the house to eat together with Chief Taofiloa and his wife. It took me time to get comfortable with seeing my age mates and older children of the family sitting on the opposite side of the house serving me, and only eating after I was finished. Before we left California, my father told us that he would come with my younger brother, Ethan, to join us in Falefa two weeks after our arrival. One day after five weeks had passed, and they still had not arrived, nor had we received any news from them, I was sitting in the main large open sided *fale* writing a letter home, wondering when or if my father and brother would ever come; when suddenly, a taxi arrives. I hear members of my Taofiloa family excitedly shouting, "Sim! Talofa lava!" Sitting there, not knowing what to do, someone barks at me, "Your father has arrived! You are his son. Show him respect! Fetch him some cold water to drink."

My father's arrival represented a change in relational contexts. Consequently, after so many days of fulfilling a role as honored guest, I suddenly needed to switch to the new role of subservient son to my father, a role that I had never played before. That was my first lesson in the flexibility of role-playing in Samoa, the sense of a person at any time only performing one part of their identity, and needing to be ready to play a different part at a moments notice.

Example 2: This example derives from my second visit to Samoa in 1980 when I stayed with the Taofiloa family for five months. In this example, I learn a lesson about the dynamic of context–based mutability of identity and role when I arbitrarily decide to shift my role without any relational contextual reason to do so. I had been with the family for about a month, during which I was consistently being treated as an honored guest during family meal time, that is, sitting cross legged on the mat on the front side of the house being served and eating first together with the elder parents, Chief Taofiloa and his wife, Fu'a, while others waited upon us. Outside of meal times, I was getting involved in learning and doing many of the Samoan chores that young Samoan men do in the village, farming, fishing, etc. but felt a little strange when it came to meal time to have my age mates of the family sitting across from me serving me.

So, one day at mealtime I arbitrarily chose to sit on the mat at the backside of the house together with the servers. The family told me to go sit in the front to be served, but I told them I would stay on the back and be a server. And so I did. Not only did I miss out on some of the choice foods that I helped cook at the *umu* and was looking forward to eating, but I also found that the family was not particularly happy about my role switching. I had wrongly assumed, based on my American notions, that they would be happy to see me humbly take the server role for a change.

The role switch was not relationally and contextually motivated by a need to fulfill a cultural script. Instead, it was individualistically motivated about what I wanted to experience. More importantly, I was not considering how me playing the honored family guest role at mealtime was valuable to the Taofiloa family's reputation in the village where hospitality is a prized virtue. Because the Samoan house is open sided neighbors and passersby could see where I was sitting, and what was happening. Me sitting in the front with the chief and his wife being served heaps of food was a positive optic for the family reputation. Me sitting in the back being a server could be a negative optic for the family. In contrast, me serving my father when I was 11 would be a positive optic for the Taofiloa family, signifying that they had trained me well in their Samoan cultural ways.

Example 3: This third example reaches further than just mutability in role shifting to mutability in identity. It was 2005. I was married with two children with Faaiuga.

The previous year I had been bestowed the *matai* (*aloali'i* or prince) name of Fepulea'i from the descent group of my father-in-law's mother's side in Faia'ai village. Through my ethnographic study I learned that the investiture of a chiefly name, which derives from a founding chief of the lineage, is understood as an insertion of the spirit of that founding chief into a living person. It does not erase the prior identity of the person who did not have a title before but does add the responsibility of embodying and manifesting the spirit and chiefly wisdom of the founding chief of the lineage. Thus, when addressing a person with a chieftain name, the pronoun *oulua* (the two of you) is commonly used to respectfully be inclusive to a person, recognizing both the person themselves, and the chief that is in them. When one is called out by their *matai* name, people say it "*fa'aola le suafa*" (i.e. brings to life the name [and spirit] of the original ancestor to whom the name belongs together with the chiefly wisdom that chiefly status embodies). Yet, the separability of the chiefly identity and that of the person embodying it is well understood, which enables a person to step in and out of the chiefly role as needed, which Mead's example at the beginning of this section of the chapter illustrated.

What I had yet to learn in 2005 when I travelled to my in-laws' village of Sili, Savai'i from American Samoa to bring items for a large ceremonial exchange was how the flexibility of the system extended to temporally assuming a chieftain name other than one's own in order to fulfill a needed role. During the lead up days to the big exchange to celebrate the completion of the village's newly rebuilt Methodist church my father-in-law, Faaofonuu (the head of a large descent group), like all other congregant families and their chiefs, was busy receiving many in-law groups (*paolo*) one after the other throughout each day. Each *paolo,* numbering ten to twenty people, arrived headed by their own *alii* (high chief) and *tulāfale* (orator chief) bearing gifts of fine mats, money, boxes of tin fish, and cloth (called a si'i) to support the congregant family for the big day of the dedication. To show their appreciation, each receiving family gifted back a *faaaloaloga* (money, fine mats, and food) to the *paolo. Tulāfale* on both sides do oratory as part of the exchange, helping to articulate the relationship ties through which the exchange occurs.

My father-in-law, Faaofonuu, is an *alii* and needed his *tulāfale, Manufotu,* to speak on his behalf. However, Manufotu was absent, so my sister-in-law Sefulu's husband, Mauaava, an *alii* of his own village, was playing the substitute Manufotu for the day. I chose to play the role of videographer as that is what I was most skilled in, and, with my interest on the implications of architectural change in Samoa, I was focused on documenting how the enclosed western style house they were using for these mini exchanges appeared to be a hindrance to the flow in and out of people and gifts of the exchange. I was not focused on the oratory or realizing that Mauaava was playing the orator role of Manufotu. Suddenly, a family member asked me to go be Manufotu, because my fellow son-in-law of the family needed a break. I protested that Fepulea'i was an *alii* name, thus inappropriate for me to do the oratory, and furthermore I never had played that role before, and was not familiar with making those speeches. But they told me it was not difficult oratory, and I could do it because I will be Manufotu. So, I did it. I became Manufotu for about 30 minutes and performed the routine Samoan oratory satisfactorily enough without fully understanding of how I did it—again the Samoan focus on the play, not the players.

Samoan Principles of *Tautua* (Service) and *Tausiga* (Care)

There is a common Samoan saying, "*O le ala o le pule o le tautua*", which means, "The way to authority is through service". *Tautua* (service to community, and the greater good, through a hierarchal structure) is a pronounced principle in Samoan enculturation that one hears repeated often and even found on signs at houses in villages where the *aumaga* (untitled men's group) meet to do various services for their village. *Tautua* is one of the important criteria examined when considering a person's worthiness to receive a *matai* title of the family and village. I was told the reason I was chosen to receive the Fepulea'i chieftain name was because of my *tautua* to the family, most specifically in the form of funding and helping build a new meeting house of the family.

Useful understandings about *tautua* start with breaking the word into its component morphemes. *Tau* refers to agency. *Tua* refers to the rear. Hence *tautua* is the agency enacted from the rear towards reaching community cultural objectives located relatively towards the "front". For example, growing taro in the family and village farms behind the household compound, then cooking it in the *umu*, and then serving it to the chiefs meeting in the meetinghouse in the front of the village constitutes one form of *tautua*. The family chief resides in front, while family members who perform *tautua* to the chief reside in other houses to the rear of the chief's residence.

At the same time, the chief performs *tautua* to the collective of the village council that meets in a meetinghouse in front of the chief's residence. The relational attribute of rank in the social hierarchy is thereby structured spatially on a front/back and center/periphery axis, which also facilitates the flexibility of identity and roles, as already discussed in this chapter. Dogs in this *tautua* framework form a lower status position in the hierarchy, expressed by them waiting outside the rear side of the house to be fed only after all the people have eaten. Hence, there is always a collective goal and entity higher than oneself, as well as lower than oneself, to which one performs *tautua. Tautua* is reconstituted in a globalized Samoa through the funds remitted from the overseas Samoan back to their family and village back home to support various projects and life crisis events. Flowing in the opposite direction of *tautua* is *tausiga* (care, protection, security, and sustenance). Thus, the chief receiving *tautua* is responsible to ensure *tausiga* flows back to those giving their *tautua*. This comes in providing that all those who belong to the descent group and perform *tautua* are nurtured and protected, have land to build and grow food, and have a sense of belonging and security within the group. With this Samoan enculturation framework it is now possible to examine enculturation of Samoan dogs.

Samoan Enculturation of *Canis familiaris*

Canis familiaris (domesticated dogs) have lived in close association with Samoan people in the Samoan Islands for over a thousand years, adapting and learning valued roles to play as the less dominant species in an evolving inter-species Samoan culture. It is interesting to contrast the fact that in some other Polynesian societies, such as in Tahiti, or in Hawaii dogs became a favored food to eat, but in Samoa this was not the case (Titcomb 1969). A possible reason for this difference could be that Samoan people observed their dogs' canine social intelligence and empathic communication abilities and the types of valued roles dogs could learn to play based on these abilities. This value outweighed any perceived dietary benefits from eating the dogs. In any case, they already had enough other protein sources from the sea, and their other domesticates, such as pigs and chickens.

My description of interspecies culture of dogs and humans in Samoa contrasts and draws similarities to other ethnographic cases presented in this volume. For example, my observations of how Samoans treat their dogs generally seems kinder than how Carucci (this volume, Chapter 9) and Prokosch (this volume, Chapter 3) both describe Marshallese treatment of their dogs. The same with how Lohmann (this volume, Chapter 8 describes the Asabano treat their dogs or how Morrison (this volume, Chapter 4) describes the treatment of hunting dogs on Hawaii's Big Island. At the other end of a dog–human spectrum of kindness to dogs in Pacific cultures and the extent that dogs play culturally valued roles in their interspecies cultures are the Arapesh (Dobrin, this volume, Chapter 2). In both these cultural cases dogs are considered as sentient beings with significant metaphorical and practical roles in the social life of their humans, including their kinship system. I see the interspecies culture in Samoa as lying somewhere in the middle of this spectrum.

Comparing treatment and behavior of Samoan pet dogs with pet dogs in Western societies highlights useful insights. Unlike what happens in most Western societies, Samoans do not typically take their pet household dogs for walks, especially not with leashes, nor buy either special dog food or toys for them. But similarly to Western pet dogs, Samoans dogs are given names by their humans, tend to be treated endearingly, and are often remembered fondly by their humans after they die. As lower ranking beings in Samoan household hierarchies, Samoan dogs typically behave more demure or timid than the typical Western pet dog. Western visitors in Samoa may assume this behavior results from Samoan people's harsh treatment of their dogs.

In response, I contend that this demure behavior is best interpreted as appropriate Samoan canine cultural behavior, hence a sign of successful canine enculturation, and not necessarily undue harsh or cruel treatment from their humans. I remember multiple instances where Samoan people noticed and criticized other humans for not taking good care of, or being cruel to, their dogs. Yet, Sir Peter Buck (Te Rangi Hiroa) stated some 90 years ago in the late 1920s that "The affection of a Samoan for his dog is psychological. He will allow it to starve and suffer from a curable skin disease, but he resents anyone throwing a stone at it" (Buck 1930:8). At first glance Buck's statement may seem to contradict my observation, but not necessarily. Neither one of us was systematically collecting data on treatment and attitudes towards dogs, so any generalizations should be understood only as impressions, not data supported facts. Plus, in the eight decades that lie between Buck's observation and my own there have been many cultural changes in Samoa, which can also account for the difference of impression. Also, the criticisms I heard about others not taking good care of their dogs were made to me, not to the person being criticized. To directly criticize how another non-family member treats their dog, or even their child, would probably be considered crossing a social boundary in Samoa. Regardless, I believe that having a strong, healthy, well treated pet dog reflects well upon the dog's owner within the community.

My observations of, and interviews with, Samoan villagers suggest that they recognize, value, and encourage their dogs' canine social intelligence and empathic communication skills to "read" human gestures, facial expressions, tone of voice, and so forth—canine abilities for which there is a growing body of substantiating scientific research (e.g., Goode 2007; Hare et al. 2002; Payne et al. 2015; Sanford et al. 2018; Silva and de Sousa 2011; Vanutelli et al. 2015). With that human support Samoan dogs, as members of Samoan households, adopt the ethos of learning to do useful tasks and play valued roles that serve collective household aims. In other words, their enculturation into Samoan culture promotes them to the best of their canine abilities to *tautua* to their humans in their interspecies households. As I will illustrate, Samoans do not specifically train their dogs to do any specific tasks for them through any kind of Pavlovian type conditioning. Rather it seems that dogs, as members of households, enculturate the value of *tatua* and then strive through their own volition, keen observation and other senses to both learn and to fulfill tasks that are helpful to the household. Regardless of their performance, they are given *tausiga* from their humans, that is care, protection, food, and affection.

Samoans classify their dogs into three main types determined by the role they play in their society: house (or household) dogs, (*maile nofo i fale* or stay at home dogs), for which I describe two valued roles they learn to play on the family compound; wild boar hunting dogs; and stray dogs, that is, dogs who lack a human family.

Household Dogs—Pig Corralling

Samoan village chiefly councils, made up of the head chief of every village descent group, govern the village and collectively establish and maintain village rules. Typically, in all Samoan villages, every family has its own pigs (as well as chickens, and now many of their own cows). Household pigs are typically kept in an area in the back of the family compound behind the family cook hut where they are also fed. Some villages maintain a rule that village families must keep their pigs enclosed in pens. Other villages, particularly on Savai'i Island, do not have these rules (at least did not for many of the years I was there), and people in those villages expressed that it was unkind to keep pigs locked up. Evidently, the opinion was that it is better for pigs to learn their place in the household than to be walled in. As a result, pigs are able to wander outside heir designated areas and cause damage to the well-tended grounds of the household compound—rooting up lawns, flower gardens, etc. When householders see pigs roaming and coming to the front areas of the compound they often throw rocks at the pigs to chase them back to their designated area to the rear of the household compound. A regular practice of my father-in-law, Faaofonuu, was to sleep at night with a slingshot and some rocks by his side and periodically wake up at night to check for pigs.

Household dogs learn from observation and their empathic communication that their fellow human householders do not want household pigs routing around on the household compound. I observed the household dogs effectively chasing the household pigs from the front part of the compound back to their designated area to the rear. I asked family members how the dogs were trained to do this task, and they said they were not trained; the dogs learned from observation, and decided they could play a useful role, hence, canine *tautua*.

At pig feeding time in the late afternoon, I also observed another type of pig corralling. Each Samoan household has a unique vocalization they use to call their own pigs at feeding time. Pigs learn to recognize their owner's call and come to the location typically behind the cook hut to eat. Also, the owner cuts a unique notch in his pigs' ears to identify them as his pigs. If and when pigs from a neighboring household come to eat their neighbor's food, the pig owner will chase those intruding pigs away. One day I was standing on the family compound talking with one of my in-laws when one of the household's dogs suddenly ran full speed to where the pigs were feeding in the back. I heard loud barking and pigs squealing, and saw the dog chasing three of the offending neighbor's pigs out of my household's pig eating area. My brother-in-law explained that they had never trained the dog to learn how to identify whose pigs were whose or to chase out the intruding pigs. Again, a case of dogs observing, and learning useful jobs they can do without specific training.

Household Dogs as Physical and Spiritual Guard Dogs and the Influence of the Built Environment

The role of protector or guard dog is the most common role housedogs play in Samoan households, particularly since pig corralling only exists in villages where pigs are not enclosed in pens. My Samoan interlocutors explained to me that dogs can better sense if an approaching human has good or bad intentions, and who loves and cares for living things (*alofa i mea ola*) or not, and bark at those that they sense may be an issue. As such, people rely on their dogs and their senses for cues about other people. This canine sensory ability is what I would assert ancient Samoans going into a battle were trusting when they relied on a dog (believed to be incarnating the War God Salevao) to signal to them what battle strategies to employ (Turner 1884:49).

Not once do I remember ever having a household dog bark at or get aggressive with me while walking through rural Samoan villages where the built environment is largely non-fenced yards and open sided houses, thereby giving maximum visibility. This includes the two-week walkabout on Savai'i Island in 1980 when I walked through dozens of villages. This observation stands in stark contrast to my experiences living in semi-urban environments of American Samoa where the housing tends to be predominantly enclosed walled semi-Western in style, there are also some fenced yards (though not a lot), and where there is greater human population transience.

For my last five years employed at American Samoa Community College in Mapusaga, Tutuila, I was appointed to develop a Health Communications Research and Media Program at the College's Division of Community and Natural Resources. The issue of aggressive dogs inhibiting a healthy environment in which people felt comfortable walking came up an important topic to discuss. Many people walking in urbanized areas feared being attacked by dogs when walking. The custom is to have a handful of small rocks ready to throw at any threatening dog. However, just the action of bending over to just pick up a rock can usually work to ward off aggressive dog attacks of pedestrians.

While writing this chapter I was pondering the observed differences in behaviors of *Canis familiaris* between more urban and semi-urban environments of either of the two Samoas, and more rural Samoan environments. Based on my observations I have come to hypothesize that more enclosed walled and fenced built environments co-relate with more aggressive territorial dog behavior, and with greater human transiency, and thereby less intra-community familiarity and mutual trust and understanding between peoples. Vice versa, the open sided and unfenced built environments of Samoan rural villages associate with less aggressive and territorial dog behavior, less human transience, and greater intra-community familiarity and levels of mutual trust. The cause and effect relationship within these associations is circular. The transience of human populations in more urban areas reduces familiarity and trust, which stimulates a desire for more firm physical boundaries, e.g. walled houses and fences, which in turn also further reduces familiarity between neighbors. All of this is sensed by the dogs and affects their behavior. Stray dogs are also a product of the human transiency that is more common in urban areas, which I will be addressing in a later section of the chapter.

It may also be pointed out that open unwalled house sides and unfenced yards enhances interior/exterior visual and physical interaction and allow dogs (as well as people) to observe and learn from people inside the house, even while they themselves may be on the outside, and vice versa. The fluid indoor/outdoor flow facilitates enhanced enculturation and learning processes for both dogs and their humans.

I decided to test out this hypothesis by calling my Samoan friend, Talanoa Lesātele, currently a practicing Samoan architect based in Cambridge, but who grew up in Samoan villages. When I called him, I told him about my observations of differing dog behaviors between rural Samoan villages and more transient semi-urban villages. I did not tell him my hypothesis but just asked him why he thought that might be. He responded with an explanation that supported my hypothesis.

Furthermore, as Talanoa pointed out, walls and fences express not only humans increased desire for privacy, but an increased feeling of territoriality, and reduced interpersonal trust, sociality, and familiarity. Here it seems that dogs use their empathy and social intelligence to read and absorb this difference in human emotions and mental state as embodied in, and created by, the built environment. Dogs behave differently according to these differing social and physical contexts. For instance, walls and fences can engender a dog to feel alienated and cut off socially from the rest of the household to which it belongs.

Dogs for Hunting Wild Boars

My primary source for understanding Samoan wild boar hunting dogs (*Maile Tuli Pua'a 'Ai Vao*) derives from ethnographic videos shot by Faafetai Saili, one of my students of an Ethnographic Video Methods course I taught in 1999 at American Samoa Community College. She followed with video camera in hand several wild boar hunting expeditions in remote areas of Tutuila Island, American Samoa.

Since wild boar hunting requires traversing long distances in rough interior terrain, it is particularly tiring for hunting dogs. They are also at greater risk of being gored by a boar's tusk than their human counter parts. Hence becoming a hunting dog requires these dogs to be strong, courageous, and altruistic. I learned that Samoans who like to go wild boar hunting discern which of their dogs appear to be the strongest, most aggressive, loyal, and courageous to become their hunting dogs. On the whole, Samoan wild boar hunters highly value their hunting dogs because these dogs play an essential role toward a hunt's success. Also, since a successful hunt requires cooperation and communication between human and dog, a close one-on-one relationship forms between the hunting dogs and their humans. This relationship is also reinforced by the hunter consistently being the person who takes special care of his dog, for example feeding the dog scraps of raw meat to help keep the dog in top physical condition, as opposed to just left over scraps of left-over food after human household members eat. However, quality of care of hunting dogs in Samoa varies. One Samoan woman in Savai'i told me that some of the hunters are not kind to their hunting dogs and even forget to feed them after they come back from a hunting trip. However, more direct fieldwork and research would be needed to get a better idea about the human/canine relationship in Samoan boar hunting, and to learn how common wild boar hunting is in either American Samoa and Samoa, as well as how many wild boars there are, and how many dogs in Samoa are hunting dogs.

Dogs' Altruism and Guarding Against Evil Spirits

I wish to recount here a story I learned from my wife, Faaiuga, about an experience her father had with a group of his dogs. The story reveals insights about Samoan beliefs about dogs, not only as physical protectors of their humans, but also as spiritual protectors.

The story takes place back in the 1970s when Fauiuga's older brother was just a toddler. He had become sick and was staying overnight with his mother at the hospital in Tuasivi, about 20 miles down the coast from the family's village of Sili in Savaiʻi. One afternoon, Poe, (my father-in-law's given name), decided he needed to bring some food to the hospital for his wife and young son, so he prepared the food, and because there was no other form of transportation, prepared to walk the 20 miles (32 km) to the hospital. However, it was approaching night, so he decided to bring his six hunting dogs with him for extra protection, especially against the evil spirits that lurked dangerously in the dark and in the bush between villages. Once he departed, he noticed that several of his household (not hunting) dogs had joined him and the hunting dogs. However, Poe did not feel that household dogs had the strength and will to fend off the spiritual dangers that lurk at night in the bush between villages. Thus, he forced them to return home and stay there.

The fact that the household dogs joined them of their own volition suggests perhaps that they empathically understood the risks and dangers involved, and altruistically wished to join the walk in order to help, despite the danger. The fact that Poe would not allow them to come with him conveys the idea that they were not considered strong and tough enough for the job, perhaps for the same reason that only the strongest and bravest dogs become hunting dogs in Samoa.

After several hours of walking in the middle of the night, the hunting dogs started howling, running out into the bush and back to the road, and running in circles around Poe. According to the story, this behavior was interpreted as the dogs fighting evil spirits that were trying to hurt Poe. When Poe was about halfway to Tuasivi he saw a glowing light on a hill where there was a small house with a lighted kerosene lantern in it. He walked there with the dogs and requested to take refuge there with his dogs for the night. He slept in the house with the family, but the dogs continued to run around outside, howling and trying to fight the evil spirits that were still on the prowl until the howling stopped. The next morning, Poe went to collect his dogs to continue the walk to the hospital, but found all six dogs were dead with their eyes poked out. The interpretation given was that the spirits had killed the dogs and poked out their eyes. They had altruistically lost their lives in their heroic and altruistic acts to protect their human. Poe buried his hunting dogs near the house, and told his daughter, my wife, about the story one day when they passed the place on the bus, also telling her how much he missed and was sorry about his dogs who lost their lives trying to protect him; he promised himself that he would never again attempt to walk somewhere distant at night. Could these dogs be considered Samoan "spirit dogs"?

Wayfinding with "Spirit Dogs" During my Savai'i Walkabout

Clues for solving the mystery of the two "spirit dogs" mentioned at the beginning of the chapter are embedded in the chapter's previous sections. Now it is time see what those clues bring to light as I retell the story of my four-day encounter with the "spirit dogs" as part of my two weeks' walkabout on Savai'i Island. This solo walking circumnavigation of Savai'i was a wayfinding, a space to navigate cultural differences I was experiencing between my primary enculturation in America and my secondary enculturation in Samoa. This chapter section applies "an embodied, intimate, and relational model for doing critical autoethnography". It is itself a wayfinding in which the author is both:

> the source of not only the story of experience, but also the role of the storyteller or interpreter of experience [and] a bridge between traditional knowledges and the European interpreters who have long studied and sought to frame it in western ways for western purposes
> (Iosefo, Jones, and Harris 2020:20).

Through an interspecies enculturation framework, I find a path to bridge the gap between the traditional Samoan knowledge, embodied in the woman's metaphysical explanation of the "spirit dogs" not really being dogs, and any Western, science-based explanations.

Lead Up to my Solo Circumnavigation Walk around Savai'i Island

When I returned to Samoa in 1980 when I was 19 to stay five months I wanted to reconnect with my Samoan family in Falefa and continue my secondary Samoan enculturation, including becoming proficient in Samoan language, and in doing all the various types of work young Samoan men in the village do. I also had designed a "field" research project to do in Samoa that I had designed my second quarter at the University in California. A journal entry written on completing my first month in Falefa highlights how my enculturation over the previous eight years in America had made me more individualistic, and thereby less well adjusted for living as a member of a Samoan family:

> Ve'a [my older Samoan brother] goes fishing every day and also at night. I ask him if I can go but he always says no. I figure I must be a burden and less chance of good fishing with me on board. Perhaps, he really enjoys doing it alone. … The best thing for me to do is to take it easy, lie back, live day-to-day, enjoy the satisfaction of love in this close communal living, the honor and respect given me, and making friends. I am learning a lot in this way, through my conversations with people. [However] it's impossible to leave this family for a while and

> live with another family without insult and trouble. I really like this family. However, I would like the experience of living in another family and getting to know other people
> (Micah Van der Ryn, Unpublished Journal, July 15, 1980).

Even though I was picking up the language rapidly and learning to do a lot of daily Samoan village work, including some fishing and farming, etc., I still felt I had too much "non-productive" time, and felt restless and constrained. After three months in Falefa I became determined to take the ferry across to Savai'i, Samoa's larger, but less commercial island to do a solo circumnavigation walk around the entire island with its roughly two hundred mile circumference—what I imagined to be a kind of Samoan "walkabout"—a traditional male rite of passage in Australia, but certainly not a part of Samoan culture. The closest male rite of passage in Samoan culture is the *tatau*, traditional male body tattoo ritual, which I would some years later make a documentary film about.

My plans went against the Samoan ethos of the collective pursuing group goals and protecting group reputation. Furthermore, my going off alone was not in accord with the Taofiloa's being my Samoan visa sponsor. They warned me that walking alone on the road was dangerous—people would throw stones at me, they said. That only made me more determined to go and prove them wrong. The elder parents told me to wait another month and I could go with them to Savai'i to attend the church meeting there. However, I did not wish to postpone a month, and attending a church conference was not my purpose. My Samoan brothers and sisters tried to help the parents understand about cultural differences, telling them that despite my efforts to follow Samoan ways I was still an American who had such crazy ideas as going off to Samoa's largest island to walk around it alone.

The First Half of the Circumnavigation without the Spirit Dogs

Landing at the wharf in Saleologa, I began my solo walkabout two days later headed counter clockwise around the island. My journal states, "Walking is very satisfying to me. It gives me time to think, I see more; it is strengthening and peaceful." I mostly followed my Falefa family's advice to go to the village church minister's house to ask for lodging at night. The advice stems from the fact that church ministers are well cared for and fed by their congregations, and, as a rule, are from outside the village, thus politically neutral. Sometimes, I accepted invitations from non-minister families to stay with them, which never caused a problem. A few times young Samoan women saw me walking on the road and ran to meet me on the road to ask me to stay at their house, but I declined because it was the middle of the day, and I was concerned about complications occurring.

In the late afternoon of my eighth day, I was halfway through my circumnavigation of the island as I reached beautiful Falealupo Village at Savai'i's most western point. Many Samoan fale with their dome shaped thatched roofs raised on posts above volcanic rock foundations stretched down Falealupo's long, broad white sand beach. In most other villages I saw more of a mix of traditional structures, and semi-newer type, rectangular shaped houses with hip roofs. A young man I met as I entered the village told me the village council had restricted building of the newer styles houses that used imported building materials. I asked him where the church minister's house was and found it. The minister's wife was there with two daughters but the minister was away. They had me sleep in their long traditional *faleāfolau* (Samoan long house) in front of a more enclosed cement house where they lived. I spent two days and nights in Falealupo, enjoying eating taro, fresh fish and lobster with coconut cream, watching the sun set into the sea, and sitting in the white sand on the moonlit night learning Samoan love songs from a young Samoan woman while children danced nearby before departing in the morning of the third day.

Two Samoan "Spirit Dogs" Join me on my Savai'i Walkabout

With the days slipping by, and only being halfway through my circumnavigation, I was set on walking at least ten miles that day. Within an hour, I was walking down the scenic coastline towards Tufutafoe Village when I spotted a man spear fishing in the lagoon, then saw a family collecting sand and rocks on the beach for a building project. Soon a man of that family, named Gise, intercepted me on the road to invite me to their house for a rest. I declined the offer, but his kind persistence won me over. Some hours later, after a swim in the sea, and a nap, the family suggested that since it was getting late, I should spend the night and leave early the next morning. I felt an indescribable spirituality emanating from the family and the place that is hard to describe. I later learned the pre-Christian Samoan belief that this was the place the souls of Samoans who had died came to before descending to Polutu (the Samoan underworld), which was entered through a whirlpool called Fafā Sau Alii just off the point near Tufutafoe. Perhaps that had something to do with the intangible sense of spirit I felt there.

The next day Gise woke me up at 5 AM as I had requested. We bid our farewells and I continued my walk in the pitch black of a moonless, pre-dawn morning with only the stars to guide me. An hour later rays of the rising sun filtering through the canopy of trees above illuminated, to my surprise, two healthy dogs walking silently with me in a stately fashion—a long-haired brown one ten feet in front of me, and a smaller, white one ten feet behind. Their appearance and behavior were a total mystery. I could not remember ever seeing these dogs before, let alone having interacted with them in any way that would initiate a bond that they would join me. I thought they possibly belonged to my host family of Gise in Tufutafoe, but I was only guessing.

My main concern was that these dogs needed to return home to their owner where they belonged. I did not wish to be responsible for them, and my aim was to do a solo walkabout unaccompanied by anyone, including dogs. I tried in vain to try to get them to leave me. I threw rocks towards them on the road and yelled "Halu!" (the unique way for Samoans to tell only dogs to "go away!"), but that did not work.

I did not wish to turn around and walk back with them to Tufutafoe because, given my schedule, I needed to keep moving forward. Thirty minutes later a large truck came by driving in the direction of Tufutafoe. The driver stopped to ask me about the beautiful dogs. I explained I thought they must belong to Gise in Tufutafoe, and the driver promised to bring them back to Gise in Tufutafoe, so I agreed. A man jumped out from the truck and caught the white dog while the brown one hid in the bushes. Then the white dog nipped the hand of the man holding him in the back of the truck so the man would release him. The white dog then jumped out of the truck to rejoin me on the road. The impressed driver shrugged his shoulders and drove off leaving the two dogs and I to continue our journey in the same formation—brown dog in front, white dog behind. I could not understand the determination of the dogs to stay with me, despite me never having had any bond I knew of with them. My diary at the time reads:

> I tried without success to send them back home. However, their unshakable determination to stick with me and protect me finally won over my heart for those dogs who acted with much love and protection for me.
>
> (Micah Van der Ryn, Journal Entry, July 1980)

A woman I met on the road several hours later admired and made enquiries about the dogs. I explained their mysterious appearance, and my failed efforts to rid myself of them, to which she replied, "*E le'o ni maile lenā, 'ae o agaga faapitoa ua omai fesoasoani ia te oe.*" (Those are not dogs, but rather special spirits that have come to help you). I thought, "They sure look, feel, and breathe like dogs, yet they are not really dogs—how can that be?" Her explanation thereby doubled the mystery to be solved and increased the sense of a cultural chasm I would need to address one way or the other.

Solving the Mystery

Investigating the woman's explanation from both etic and emic perspectives provides the basis for creating a bridge. From the etic side, there are scientific observations that abnormal animal behaviors can be interpreted as those animals being something other than what they appear to be. For example, Lohmann (2003:202) mentions an Asabano boy (in Papua New Guinea) who interprets a bandicoot he sees in the daytime as an incarnation of an evil witch; bandicoots are usually only seen at night. Forth (2022:495) describes how turtles and coconut crabs seen in inland forests of Flores Island, Indonesia may be interpreted as embodied forest spirits because turtles and coconut crabs normally are found at the sea, not inland forests. Note that the rarity of occurrence is defined as a characteristic of the supernatural explanations. Likewise, the behavior of the "spirit dogs" I encountered was rare for domesticated dogs, thus, they were not really dogs, but special spirits. But rarity or abnormality of the behavior does not explain the reason for it. For unusual behaviors of non-domesticated animals, such as bandicoots or turtles, I would suggest that there could be ecological reasons.

No quantitative data for the frequency of "spirit dogs" in Samoa exist. If they were common, then perhaps there would be a direct Samoan term to reference the behavior. The term "spirit dog" I only use for the sake of this chapter, and in all my many years in Samoa, I never heard of any other occurrence of it. I asked my wife about the story of her father and the six hunting dogs killed by evil spirits at night, if those dogs were "spirit dogs". Her reply, "No they weren't, because if they were spirit dogs, they would not have died." When I asked her about the "spirit dogs" that walked with me, she was quick to say, "No, those were not dogs, but spirits." I then decided to call Talavou, a friend who now lives and works in Oklahoma, but who I first met during the walk around Savai'i in 1980 and had directly witnessed me with the dogs. I asked him what he thought about the dogs. He responded that he didn't want to tell me at the time, but that he thought they were not really dogs, but rather angels sent from God. I then spoke with my wife's sister about it and asked her if she had heard of spirit dogs; she said, she had, and also "spirit cats"! I told her the story of the mystery dogs, and asked her how would she know if those dogs were "spirit dogs" or not. She replied, "They had to be 'spirit dogs' because no normal dog will leave his or her owner, the way those dogs did!"—a statement that directly marks agreement about abnormal or rare occurrences as generative of spiritual or supernatural explanations in some cultures.

The "supernatural" idea of "spirit dogs" runs against scientific materialist thought, but matches with the ontology expressed in many facets of Samoan culture that I have researched, experienced, or read about. This includes Samoan spirit possession of humans (*ma'i 'aitu*—ghost sickness), and in the concept in the *matai* system that the ancestral spirit of a founding chief is invested in the living person as part of the title investiture, and in Samoan legends, where various different spirits are incarnated in special rocks, trees, and animals. Those legends are not seen as old make-believe folk tales but rather as a present day living reality of their inhabited landscapes. Perhaps the most contiguous is the pre-Christian Samoan belief and practice of the Samoan war god Salevao, incarnated as a white dog who assisted Samoan war parties in their battle strategies.

The fluidity of roles and identities made possible by this ontological view of being is consistent with the malleability of identities and roles I described as a principle of an interspecies Samoan enculturation framework. In other words, a being, human or dog, can, on a moment's notice, become a different kind of being, for example, a "spirit dog", in order to fulfill a perceived needed role in a cultural script prescribed by a specific context. In this case, the context was me walking out alone in the pitch black of a moonless predawn morning leaving behind a kind and loving host family in Tufutafoe. From a Samoan cultural point of view, my action was dangerous and required at least one or more other fellow beings to join me. The dogs saw that Gise could not join me, so either the dogs empathically received a message from Gise to join me, or they directly perceived it would be a useful and valued role to play to become my "spirit dogs". Thus, I propose that the two Samoan "spirit dogs" that joined me on my Savai'i walkabout were a special and rare outcome of their interspecies enculturation within Gise's household. That enculturation most likely involved the dogs' humans supporting and encouraging them to develop and fully use their empathic communication and social intelligence.

At this junction in the story, I wish to address a comment from on an anonymous reviewer of an earlier draft of this chapter. The reviewer stated that in their personal experience, "across Polynesia it is not uncommon for Polynesian dogs to follow people around islands". The reviewer also states that they assume this "to be a learned behavior that typically results in the dog being led to a person's home where it will eventually receive food scraps". The reviewer describes how they personally experienced a dog following them and their crew "for days across a mountainous Polynesian island without being fed or encouraged" and states "it is possible that someone raised in Polynesia would call this a 'spirit dog' and reward it for its protection, but from the dog's point of view it received a reward for following a human".

First, not all Polynesia is culturally the same, and that includes how dogs are treated (e.g. the neighboring Tongans enjoy eating dog meat), so generalizing about Polynesians and Polynesian dogs may be an issue to dissect. In any case, it would be helpful to know more details about both where in Polynesia the reviewer observed this canine behavior, and more details about both the dog that followed them for several days with no encouragement, with all the surrounding circumstances.

As far as applying the reviewer's comment as an optional explanation for the "spirit dogs" I experienced in Samoa, I can say first that I have already substantiated that "spirit dogs" are rare, not common, in Samoa and that rarity is why Samoans interpret them as being non-dog spirit beings. Secondly, the two "spirit dogs" that joined me on my circumnavigation of Savai'i appeared from the start to be well fed, and cared for by their humans, so it is hard to imagine that they joined me and stayed with me for multiple days with the hopes of being fed food scraps.

The Second Half of the Circumnavigation and Losing the "Spirit Dogs"

While hearing the woman I met on the road in Savai'i tell me that the dogs were not actually dogs, but rather special spirits conflicted with my more scientific materialist education, it did relieve me of some of the guilt and stress for not immediately finding and returning the dogs to whomever they belonged. I figured I would just go with the flow.

By afternoon that first day walking with the "spirit dogs", we reached Faia'ai Village where I met a young man named Talavou, who led me and the dogs down a steep cliff to a beautiful, coconut tree fringed white sand beach where I swam while the dogs dutifully guarded me. Talavou told me if he did not have to join the *aumaga* (untitled men's group of the village) on a village work project the next day, he would like to join our walk to Sili Village, where I was planning to stop for a couple of days in my circumnavigation of the Savaii Island.

By evening the dogs and I reached Sagone Village, where one of the high chiefs invited us to stay in their home. The family let the dogs enter the house, fed them, and the dogs slept a few feet away from me on the mat where I noted how the dogs dutifully watched over me. The family did not ask me questions about the dogs, nor give me any explanation about them as being something other than dogs, but did seem to give the dogs special reverence.

By mid-afternoon the next day the dogs and I, with them in their usual walking formation—brown dog ten feet ahead of me, white dog ten feet behind—were about a mile from Taga village when four young men clearing land to plant crops saw us and ran to meet us on the road with a rope and their machetes. One of them began persistently asking me to give him one of the dogs. Since I had two dogs, I could spare one. Flustered, I told him I could not give him any dogs, as they were not my dogs to give away. My journal reads:

> The dog desperately sought my protection, but instead I trusted the dog could keep itself from being caught. The boys wanted me to hold the dog so they could get the rope around its' neck. I was very nervous. I wanted the dog to run ahead and escape and rejoin me later, but unfortunately that white dog stayed behind me, and the boys got between the dog and me, and then the one that made a lasso from the rope began chasing it.
>
> (Micah Van der Ryn, Journal Entry, July 1980)

Upset by the man's action and the loss of the white dog, I continued walking with only the brown dog in front. A hundred yards ahead I saw a family preparing a large *umu* (rock ground oven), They had seen the man chasing the white dog, knew his name, and invited me to wait there for him to return, hopefully with the dog. They also invited me to share some of the *umu* food with me once the *umu* was cooked. I politely declined the kind offer and kept walking with the brown dog leading the way into Taga Village proper, where the first person I met was the wife of the man trying to lasso the white dog. She invited me to rest at their house and await her husband's return, but I declined and kept walking with the brown dog.

By the time we reached the eastern end of the village it was past 3pm and blazing hot. That's when an attractive young woman joined us, asking where I was going. I told her we were walking to Sili (about 13 miles away) and about the white dog being chased by a man of Taga. She told me it was hot and late, and the next nine–mile stretch of road had no houses, or villages, and was called, "*O le Ala i le Oti*" (the way or road to death). She told me I should stay the night in Taga, at her and her family's house. We could visit the other family with the man with the lasso that evening to see if he had the white dog or not. Feeling her concern, compassion, and sensibility, I agreed, and followed her to her house where I met her father, her brother and brother's wife. That night the girl and I walked to the other side of the village to meet and talk to the family that could possibly have the white dog. The young man who had chased it told me he never caught the dog and believed it must have run all the way back to Tufutafoe. We also searched the premises to be sure the dog was not actually there. As I hung around a bit that evening with the girl, I felt I was becoming fond of her.

The next day, the girl, her brother, his wife, the brown dog and I walked out to the Taga blowholes where we threw coconuts in the holes to see them get shot up a hundred feet into the sky each time a wave entered the lava tube. As we were walking back towards the road the brother asking me if I could give him the brown dog. I told him I could not because I needed to return the dog to Tufutafoe. He agreed to do that for me, but I felt he was just saying that to make me agree to leave the dog. So, I felt unsettled about his request while also thinking maybe his sister would like to walk with me, but that idea I knew was far-fetched. At the same time, I saw the brown dog following around the brother's wife instead of me. Clearly, this signified to me that whatever "spell", or reason in the dogs' mind for following me from Tufutafoe, that is, becoming my "spirit dogs" was now broken, but why? Was it the absence of the white dog, or my attraction to the girl, or was it the brown dog already sensing what was about to happen?

As we reached the main road, I was pondering my next move when we suddenly saw a fast-moving truck headed towards us, going east, the direction I planned to go. Before I knew what was happening, the family waved the truck driver to stop and gestured me to take the lift at least to just the next village. Following their orders, I found myself a minute later sitting in the back of the truck speeding down *O le Ala i le Oti.* (the Road to Death), feeling remorseful as I viewed the family with the girl and the brown dog receding in the distance behind the cloud of dust turned up from the truck. Not only did I feel irresponsible about not getting the dogs back to Tufutafoe, but I had also broken my vow to only circumnavigate the island on foot.

Ten miles later, I jumped back down from the truck to the road, to find the road leading to the inland village of Sili, where I planned to stay a night with the family of a man I had met in Saleologa where I had begun my circumnavigation. Little did I know that 18 years later I would be returning to Sili with Faaiuga, my future wife, to meet her family and see how the village had changed in 18 years. A few days later I was standing on the ferry's stern headed across the straight back to Upolu Island, and Falefa Village, to rejoin my Samoan family of Taofiloa.

Returning to Falefa: Going Astray, and Stray Dogs

Hopping off the bus in Falefa I was elated to return to my Samoan family and share with them stories of my Savaiʻi trip. Upon entering the family compound, however, I did not sense the usual familiar warm reception. As I later learned, my going off alone to walk around Savaiʻi had put the family reputation in the community at risk in two ways. First, the family was my Western Samoa government visa sponsor with whom I was expected to live throughout my five-month stay. Second, going away from the family for no other purpose than to walk around the other island did not make cultural sense. In fact, it threatened the family's reputation in the village. Perhaps my individualistically motivated leave of absence had spurred some negative village gossip, that was injurious to the family. I now better understand the Taofiloa family not wanting me to go to Savaiʻi alone, why the two dogs in Tufutafoe may have joined me, why Talavou, the man I met in Faiaʻai Village wished to join me on the walk, and why the family in Taga village waved down the truck to take me on the next stretch of road, even though I was planning to walk that stretch of road called "The Road of Death" alone.

The Taofiloa family and I decided the best solution was for me to take the ferry over to American Samoa to visit Merelini, the daughter of the family who had taken care of me when I lived in Falefa when I was eleven. The family in Falefa had not seen her for several years, but had learned that she, her Samoan dentist husband and their baby daughter had recently moved from Australia to American Samoa where her husband had gained employment as a dentist at the hospital. During my visit I explained my ruptured relationship due to enculturation into the more individualistic American Society. She recognized the cultural difference issue at stake and told me she would go visit the family in Falefa to explain and help mend that rupture, which she did. I returned to Falefa three years later in 1980 when I sailed to Samoa on a sailboat.

While in American Samoa, I gained my first experience with Samoan stray dogs, who hold low social status among the Samoan canines by virtue of their lack of identity as members of a Samoan household and thus generally receive poor treatment in the society. My going astray with my family in Falefa preconditioned me to empathize with these stray dogs in American Samoa. My observations indicate that stray dogs are not found in Samoan villages, but only in urban zones of the Samoan islands where human transitory residence often leads to dogs being left behind as strays when people have to move, and these strays may then also reproduce new stray dogs. Without protection, care, and food from their human counterparts as members of Samoan households, these low status stray dogs are mangier, and much more timid than dogs living as Samoan household members. These social facts also help me to confirm that the "spirit dogs" that joined me on my Savai'i walkabout were not stray dogs looking for a handout.

The lack of household membership status of stray dogs also implies, especially for second or third generation stray status, that they miss the socialization process that household dogs acquire, and by which they learn to play valued roles within their households including guarding, corralling, hunting, or becoming "spirit dogs".

Conclusion

Over millennia an interspecies Samoan culture evolved in these lush, volcanic high islands, later nicknamed "the cradle of Polynesia" in which *Canis familiaris* became a favored household pet learning to play valued roles within Samoan society as low status members of Samoan hierarchal family households, which are component parts of Samoan descent groups that both comprise and interlink Samoan village communities. Over the centuries Samoan people observed, and increased their awareness of Samoan dogs' hyper sensory, social, and empathic capabilities, which they interacted with in ways to support them learning to develop useful and valued roles in the society, such as guard dogs, pig corralling, wild boar hunting, and "spirit dogs".

My examination of the interspecies canine–human culture of Samoa largely focuses on Samoan enculturation by which both humans and dogs learn to play valued roles in the society. I employ autoethnographic methods to solve the mystery of the Samoan "spirit dogs". While I do not feel that the mystery of the two "spirit dogs" that journeyed with me for three days on my walk around Savai'i Island in 1980 is fully solved, I do see that it opened new avenues for interspecies cultural anthropology research.

Writing this chapter has also augmented my in-depth research on the relationship between architectural changes in Samoan villages and the intangible dimensions of Samoan socio-cultural change, since, as I reported in this chapter, those architectural changes appear to also affect dogs' learned behaviors within the culture. Clearly, widening our view of human culture to include other species is a field that is likely to continue to grow in anthropology.

References

Adams, Tony, Stacy Holman Jones, and Carolyn Ellis. 2015. *Autoethnography*. Oxford: Oxford University Press.

Buck, Peter H. (Te Rangi Hiroa). 1930. *Samoan material culture*. Honolulu: Bishop Museum.

Churchward, William B. 1887. *My consulate in Samoa: Record of four years sojourn*. London: Bentley.

Forth, Gregory. 2022. Mountain turtles and giant crabs: Cosmological implications and supernatural understandings of rare creatures on an Eastern Indonesian island. *Anthrozoös* 35(4): 495–513.

Goode, David. 2007. *Playing with my dog Katie: An ethnomethodological study of dog–human interaction*. West Lafayette, IN: Purdue University Press.

Greig, Karen et al. 2018. Complex history of dog (Canis familiaris) origins and translocations in the Pacific revealed by ancient mitogenomes. *Scientific reports* 8(1): 1–9.

Hare, Brian, Michelle Brown, Christina Williamson, and Michael Tomasello. 2002. The domestication of social cognition in dogs. *Science* 298(5598): 1634–1636.

Iosefo, Fetaui, Stacy Holman Jones, and Anne Harris, eds. 2020. *Wayfinding and critical autoethnography*. London: Routledge.

Lohmann, Roger Ivar. 2003. Supernatural encounters of the Asabano in two traditions and three states of consciousness. In *Dream travelers: Sleep experiences and culture in the western Pacific*. Roger Ivar Lohmann, ed. Pp. 188–210. New York: Palgrave Macmillan.

Mead, Margaret. 1937. The Samoans. In *Cooperation and competition among primitive peoples*. Margaret Mead, ed. Pp. 282–312. New York: McGraw Hill.

Payne, Elyssa, Pauleen C. Bennett, and Paul D. McGreevy. 2015. Current perspectives on attachment and bonding in the dog–human dyad. *Psychology Research and Behavior Management* 8: 71–79.

Petchy, Fiona J. 2001. Radiocarbon determinations from the Mulifanua Lapita site, Upolu, Western Samoa. *Radiocarbon* 43(1): 63–68.

Poulos, Christopher N. 2021. *Essentials of autoethnography*. Washington: American Psychological Association.

Sanford, Emily M., Emma R. Burt, and Julia E. Meyers-Manor. 2018. Timmy's in the well: Empathy and prosocial helping in dogs. *Learning and Behavior* 46(4): 374–386.

Shore, Bradd. 1982. *Sala'ilua: A Samoan mystery.* New York: Columbia University Press.

Silva, Karine, and Liliana de Sousa. 2011. Empathic dogs: Theorizing about the possible domestication of emotional contagion. *Journal of Veterinary Behavior* 6 (1): 69–70.

Smith, Anita Jane. 2002. *An archaeology of West Polynesian prehistory.* Canberra: Pandanus Books, Research School of Pacific and Asian Studies, The Australian National University.

Stair, J. B. 1897. *Old Samoa: Or, flotsam and jetsam from the Pacific Ocean.* London: Religious Tract Society.

Titcomb, Margaret. 1969. *Dog and man in the ancient Pacific, with special attention to Hawaii.* Honolulu: Bishop Museum.

Turner, George. 1884. *Samoa, a hundred years ago and long before.* https://www.gutenberg.org/ebooks/14224.

Van der Ryn, Micah. 2012. The difference walls make: Sociocultural dynamics and implications of change in Samoan architectural traditions and socio-spatial practices (1940–2006). Ph.D. Thesis, University of Auckland.

Vanutelli, Maria Elide, and Michela Balconi. 2015. Empathy and prosocial behaviours: Insights from intra-and inter-species interactions. *Rivista Internazionale di Filosofia e Psicologia* 6(1): 88–109.

11 Sniffing toward the Future

Accessing Interspecies Enculturated Dogs' Perspectives

Roger Ivar Lohmann

> Trying to understand a dog's perspective is like being an anthropologist in a foreign land—one peopled entirely by dogs. A perfect translation of every wag and woof may elude us, but simply looking closely will reveal a surprising amount. So let's look closely at what the natives do
>
> Alexandra Horowitz (2009:31).

> The goal is, briefly, to grasp the native's point of view, his relation to life, to realise *his* vision of *his* world .... We ... must study what concerns him most intimately, that is, the hold which life has on him. In each culture, the values are slightly different
>
> Bronislaw Malinowski (1984[1922]:25).

To conclude this book's island-hopping, tail-wagging, meandering sniff along the sea-and-land trails of dogs and their humans in Oceania, I want to do a full body shake—as dogs do to clear their minds for a fresh start—and tentatively stick a cold, wet nose into the question of how we might use anthropology to bring us closer to understanding *dogs'* experiences, *their* understandings of the world, and *their* perspectives. Anthropological methods, even though designed for the study of humankind, could get us closer because dogs (and people) undergo different degrees and kinds of interspecies enculturation depending on who they live among and what norms they learn. Anthropology is all about connecting with and understanding culturally different others to better understand humanity writ large; So full-on, anthropological studies focused squarely on diverse, culturally distinctive groups of dogs rather than on their human associates to better understand dog-and-humankind are possible.

As the chapters in this volume make clear, canine lives, understandings, and values differ depending on the cultures that surround and infuse them. Dogs raised in different communities have distinct preferences, training, and perspectives (Horowitz 2021). They learn through interactions with culturally distinctive people and dogs in their communities to experience, think, and act in ways that would be inappropriate—or even impede survival—in other cultures. To get at dog's capabilities, tendencies, and life-

DOI: 10.4324/9781003658696-11

worlds, like those of other culture-bearing species, we need to endeavor to "understand their cultural capacities and practices" (Fitzpatrick and Andrews 2022:1108–1109).

Dogs learn current and locally appropriate habits by observing and mimicking other dogs and humans (Coren 2004:227). For humans, and evidently for dogs too, although it is often tempting to go our own way, it feels good to do what those whom we admire and identify with do. Successful dog training, according to dog trick show performer and trainer Kyra Sundance and her canine "co-author" Chalcy (2007), requires taking the dog's point of view into account to communicate and motivate, making a fun social experience for partners of both species (see also Porter 2018). Dogs adopt practices for which they feel rewarded by their humans (Musharbash 2018:25).

We don't have to look beyond 18th century Hawai'i to find a range of diverse, socially learned, interactional human–canine traditions: some dogs were cherished pets and nursed at human breasts while others were penned livestock to be suffocated and cooked. Their meat provisioned chiefs and offerings to gods; their teeth were made into fishhooks or decorative money, and their fur lined cloaks (Titcomb 1969). Despite such diverse treatments, Katherine Luomala (1960:221) remarks, "The frequency with which early European artists showed dogs in canoes suggests that dogs perhaps always made sure that they were not abandoned". Dogs, like people, get into and stay in routines, even ones that are bad for them. They embody and enact the culturally distinctive information that they have taken on through the social interactions they have experienced with their canine and human consociates (another range of examples is found in Losey et al. 2018). Knowing that someone is a dog, one cannot say, for example, what language's words or what system of gestures he or she will recognize and act on.

Dogs experience the world and communities that they share with humans in overlapping but different ways than humans do. When it comes to dogs, there are not two hands, but four paws. Their mouth is like the human hand for grabbing, their tongue like human fingers for caressing, but these organs do not neatly map onto human bodies. Although they participate in cultures, their cultural diversity is expressed and experienced differently and less elaborately than that of humans. How the world looks to dogs is subsumed under how it smells—and the significances that dogs have learned in their societies to assign particular scents. A drug-detecting airport dog in Canada has socially learned to pay attention to different smells, and to assign different meanings to that scent palette, than her contemporary Micronesian atoll-dwelling dog. Wouldn't it be worth getting to know what these very different ways of being a dog are like?

Just as other people and their worldviews are never entirely inscrutable to us, neither are the experiences of other animals whom we are privileged to know (Mezzenzana 2023:200). Anthropology circumscribes some of the best methods to access worldviews, with particular attention to unfamiliar ones. *Anthropology* means "the study of humankind". But humankind is neither separate from nor independent of everything else. Humans evolved from and may evolve into other species—or express cultural information as genetically engineered chimeras (Alter 2007). We are enmeshed parts of ecological communities. By studying other living things and their perspectives, we are reaching out to more inclusive aspects of our greater selves.

Scientific evidence indicates that dogs and humans, like the rest of life on Earth, are literal cousins (Dawkins and Wong 2016). The common ancestor shared by your dog cousin and you was certainly a single species of mammal. Without that joint ancestor, and without the diverse culturally infused interactions that began between its human and canine umpteenth-great-grandchildren, we would not any of us be what or who we are. I daresay most humans do not learn in our ethnic cultures to regard or treat dogs as the cousins they are, but the range of cultural models of the relationship is broad and worth knowing.

For example, ethnographer Deborah Bird Rose (2011:7–8) writes of an Indigenous North Australian named Old Tim, who told her that "a dog-human person" was the common ancestor of both canines and humans. "Think about it," she muses:

> to look at the face of a dog is to see your own ancestor and your contemporary kin. It is to see mothers and fathers, sisters and brothers. They still walk with us and nuzzle us with their long-nosed faces. Without them we would not be who we are, and their faces turn toward us in the knowledge that we are here together, all of us, living and dying in our related ways.

Through exercises in ethnographic empathy like Rose's poetic depiction, we comprehend how unfamiliar cultural notions shape informants' experiences and actions on the world. We can cross cultural barriers of understanding. We can peer beyond species barriers of understanding, too; even when these other understandings are strikingly different from the ones delivered by our own biological patterns, ethnic and religious predilections, or scientific research. Through trying on the experiences and worldviews of others we are also better able to see our own minds' fanciful representations of facts for what they are. That may be hard to conceptualize or face. It would, I think, be worth trying to find out by close interactive observation how particular dog groups see their humans and how that jibes (or not) with their humans' models of themselves as dog owners, partners, or relatives.

In a classic essay, Jakob von Uexküll points out that animals, be they humans, ticks, or dogs, have "perceptual worlds"—pictures of reality that are distinctive to their biology. He uses the German term *Umwelt* (which simply means "surrounding world" or "environment") to identify this species-particular picture of one's surroundings. Von Uexküll (1957:5) writes that to peek into the subjective *Umwelt* of another species:

> we must first blow, in fancy, a soap bubble around each creature to represent its own world, filled with the perceptions which it alone knows. When we ourselves then step into one of these bubbles .... A new world comes into being .... the world as it appears to the animals themselves, not as it appears to us.

We can indeed approximate and try on others' experiential worlds in this way. For an enculturated species like dogs, though, there is no single species-wide subjective *Umwelt* for us to depict.

Different dogs in different social situations will have somewhat different perceptual worlds that are made in part of culture. To know them, we need to collect and use knowledge of dogs' cultural characteristics—such as ways of thinking, interpreting, and acting shaped by their history of social learning—and how these interface with their biology. This is a job for anthropology, with its heightened sensitivity to cultural variation and methods for documenting and analyzing it accurately. It has become clear that anthropology's careful use of empathy to understand culturally different people works for studies of animal behavior, too, for "empathy can yield a better science—one that takes animals' mental lives seriously" (Webb et al. 2023:227). Practices of getting to know particular animals and describing their experiences and characteristic behaviors as they might if they could use language can convey empathy to others. This technique is used by animal rescue organizations when they post brief autobiographies on behalf of shelter dogs to garner support and entice appropriate adopters (Savvides 2013:341).

Anthropological empathy is a developed skill to understand how and why others experience and act by recreating virtual versions of their mental models of reality in researchers' minds informed by close and extended engagement (Lohmann 2011). Anthropological empathy is designed to guard against facile and uninformed projection of our own experience onto others, or worse, assuming that others have neither minds nor experiences worth knowing.

An anthropologically inspired study of dogs would involve the researcher going *and staying awhile* "in the field": where the dogs already are. If they are our living contemporaries, we have much to learn about them by being with them as they do what they do in their places and social contexts rather than relying solely on artificial laboratory or experimental settings. If they are dead, we can go where their corpses and their effects can still be found. We can infer much about how contemporary and even past dogs think and perceive from how they act and the physical results of their actions. In the latter case, the route to ancient dogs' perspectives must often be doubly inferred by considering the material record of how their humans treated and hence thought about them. This is reflected, for example, in the manner of deposit of canine corpses associated with human burials (Koungoulos et al. 2023).

As always, information about others' mental lives from behavior is more readily available from the anthropological study of living populations. There are already efforts to marry ethology's systematic behavioral data collection and characterization of animal species with anthropological ethnography's long-term engagement with others' social lives to learn more intimately and richly about the experiential worlds of nonhuman animals. For example, John Hartigan (2021) studied wild horses in Spain who are subject to ritual roundups by local human villagers. Focusing on how the horses experienced this event as a break in their ordinary lives rather than its meaning to the humans who perpetrate it, Hartigan revealed the previously unrecognized role of mares in re-establishing social groupings after the event. Beyond this, the upset these horses suffered and the value they placed on patterned distancing from one another becomes visible and poignant to us in his account. Their ability to control their social space when humans confined them in tight enclosures broke down and they violently turned upon one another. The humans struck the gathered horses to daze them and reduce their resistance to ritual shaving. After they were shaved and released into larger enclosures, we can relate to their renewed anxiety as mares attempted to locate their foals from whom they had been separated, after their scent signals had been made less distinguishable in the mash-up. Some male foals were not to be found at all. They had been killed by humans in an effort to control the wild horse population. The stallions spaced themselves apart from the fray, trying to re-establish appropriate social distance, and the mares worked to clean up the chaos and reassemble in groups with the appropriate stallion. The horses' trauma is palpable thanks to the empathy with which Hartigan interactively observes and communicates these horses' ordeal and the social order that it broke. To what extent, I wonder, were these horse behaviors cultural and characteristic of this group and not others?

No doubt the conventions and perspectives that Hartigan's Spanish horses learned from their culturally distinctive interactions with humans are quite different from those of North American stable equines studied by Kendra Coulter (2018:60); she points out that "anthropologists' ethnographic expertise provides one compelling way to study, and in turn, write about other animals" such as the horses she knows, and "the specific horse–human cultures in which they are enmeshed". A horse–human culture of this sort is, of course, what I would term an interspecies culture (this volume, Lohmann Chapter 1).

There are other examples of anthropologists applying their research tools to studying groups of other animals living in association with distinct human societies with promising results. Ethiopian hyenas and the humans they interact with come to life as historically and culturally situated individuals in Marcus Baynes-Rock's (2013) account of the last hours of a poisoning victim's life in what he reminds us is a "multispecies commons".

If I, for one, had done my year and a half of fieldwork at Duranmin, Papua New Guinea focused on the local dogs rather than humans, I would have been able to say a lot more about the lives and experiences of Asabano dogs who loved hunting, those who were home bodies, and others. I have a memory of a skinny dog hunting and eating cockroaches in an Asabano house. My thought was, "Poor thing, he is so hungry he is reduced to this! But thank goodness he has something nutritious to eat." But there was so much more I could have found out about that dog's life and experience if I had turned my anthropological training to that task. Was this an unusual and desperate measure, or simply an entertaining activity and a snack? Did other dogs do this, and if so, which ones and when? What was that dog's bug-hunting technique? What was the context of experience in which he learned to accept and value this activity? What did he and the other village dogs do all day? Which of the local dogs were close associates with whom and which if any kept their distance? I could follow the dogs around, look at the houses, the clearings between them, the pathways, gardens, and forests as the dogs traversed and inhabited them. I could have interacted with them to find out in more detail what kinds of engagements with humans they accepted and rejected, and what could be learned from them. That would have been as worthy an anthropological research project as the human-focused one that I did.

I have elsewhere characterized all field research methods in anthropology (including biological, archaeological, cultural, and linguistic topics) in terms of five categories: material observation, biological observation, behavioral observation, direct communication (interviewing), and participant-observation (Lohmann 2006). These methods put us in extended contact, either with those we study or with their things. We are guided in this effort by a scientific footing, which keeps us committed to discovering facts rather than reinforcing our own narrowness and misunderstandings. Some of these facts are not apparent from either our own native perspectives or those of our research participants. Other facts concern the content of the study group's perspectives themselves. We learn as much as we can about how the world looks to our informants and then imaginatively try on their perspectives.

When those we study are alive, participant-observation and direct communication provide some of our richest sources of information on others' experiences, deeds, and understandings. However, helpful information can also be provided by detailed observations of their behaviors, artifacts, bodies, and ecological effects. The material traces of their past enculturated thoughts and actions persist in situ long after they have died for anthropological field study. With the data so collected and analyzed, we imaginatively reconstruct for ourselves a version of the world as they understood and lived in it. Empathy is vital to such an endeavor (Russell 2010:13).

As Clinton Sanders and Arnold Arluke (2007:67) point out, "The accuracy of our understandings of the other—human or nonhuman—is grounded primarily on our sense of his or her body and behavior. The validity of this understanding is confirmed or denied by its practical outcomes". Alexandra Horowitz (2009:242–243, 258) beautifully describes using this method to understand dogs in terms of how their biology shapes their perception and thought. We humans can imagine ourselves as a dog, lower to the ground, on four legs, smelling the events of the past week. But we need to go further, for dogs are "inculcated in the human culture" in which they participate (Horowitz 2009:252). Without recognizing and hence enabling ourselves to vicariously experience cultural differences in dogs, we are left with an impoverished and monolithic understanding of canines.

We can discover dogs' cultural traits by observing the changes they have made in physical objects and settings, noting how these are different from those made by dogs in other places and times. We can observe their behaviors, attuned to locally distinctive patterns. We can communicate with them directly: not in so refined and detailed a form as we might with fellow humans, particularly those with whom we share a common language, but the degree of communication between dogs, and between dogs and humans affords impressive opportunities for mutual understanding and cooperation (Hare and Tomasello 2005). Piers Locke (2017) describes his tactile communication with a Nepali working elephant he regarded as a "nonhuman informant". Our ability to use language for ethnographic interviews with non-humans is necessarily severely constrained or nonexistent. Yet our rich communicative interactions with our dogs makes it obvious that something akin to conversing with them is feasible.

We can also do participant-observation among dogs—again not so fully as is possible with our more biologically similar fellow humans, but we can join in with dogs for extended periods in quite immersive ways. Elizabeth Marshall Thomas (2000) demonstrates that training and experience in anthropological ethnography, when turned to dogs living right under our noses, reveals astounding details about their social lives. She (Thomas 1993:viii–xi) even documents "customs" that they "emulate" that vary between populations. For example, the dogs of one household she observed, following the example of their highest ranked member, adopted the unusual practice of lying down while eating their communal meals. When one of the lower-ranking dogs moved to a new, multi-dog home where the convention was to stand while eating, she at first continued her previous household's practice, but soon conformed to local practice. Getting this kind of information requires being aware of and temporarily setting aside one's ethnocentrism and species-centrism and opening oneself to the dogs' culturally influenced point of view. "What it basically comes down to", Marc Bekoff (2007:74) writes:

> is that as humans studying other animals, we cannot totally lose our anthropocentric perspective. But we can try as hard as possible to combine the animals' viewpoints to the ways in which we study, describe, interpret, and explain their behavior.

Anthropology is all about extricating ourselves from our native assumptions by reaching out and living among others and their things. We get closer to experiencing the world as they do by putting ourselves as closely as we can into our subjects' setting. We make careful observations and engage in intimate interactions with them. In this way we often discover astonishingly foreign and enlightening ways of being. We even come to see our own ways of acting as they must appear to those we study. The provinciality and problems of our own native culture become visible to us through this different or broader perspective. The internal sense and even majesty of the foreign culture dawns on us, and the bigger reality that encompasses all cultures rises in our awareness. We understand though we do not necessarily agree. We can imagine thinking and experiencing that way, even if we do not routinely do so ourselves. The light of carefully and deeply taking others' perspectives—those that are really not entirely foreign, but those of our greater community, as humans, as animals, or as the biosphere—shines bright upon us. We form a fuller picture of ourselves and the greater reality of which we are enmeshed aspects.

Dogs are more different from humans in bodies and understandings than any two human groups are, yet members of the genus *Canis* in our own societies are, after all, *familiaris*. Although humans belong to the genus *Homo*, we are also, after all, *sapiens*: We're supposed to be smart. We can communicate rather well across species lines to dogs, despite humans' relatively limited sense of smell and dogs' relatively limited linguistic abilities. We are not so different that we cannot live among dogs and their things, communicate with them, and put ourselves in their place. We can limit our anthropocentric, ethnocentric, and egocentric projections by taking knowledge of dogs' differences and similarities into account. Intersubjectivity is an anthropological specialty that can connect us not just to other, differently enculturated people, but other, differently enculturated animals (Noske 1993).

As Elizabeth Marshall Thomas (2015:ix) remarks in her foreword to Marcus Baynes-Rock's (2015) joint ethnography of Harar humans and hyenas:

> When anthropologists study cultures that involve animals … little notice is taken of the animal side, although the animals may strongly influence the people's lives, not simply by existing but by behaviors they have chosen to adopt …. groups of the same species living in different areas behave in different ways …. when groups of … animals become involved with people, the cultural features of both participants adjust.

Thomas continues that she regrets not having studied the lions of the Kalahari when she studied the people there in the 1950s, for at that time and place, lions had learned to coexist peacefully with the !Kung-speakers there, while other lion groups hunted people. Ethnographically studying two associated species at once doubtless improves one's grasp of the perspectives of at least one of them toward the other because their attitudes are in part mutually determined.

For example, the attitudes and perspectives of named individuals of an unusual pod of orcas who routinely cooperated with human whalers come through from human memories of interacting with them (Clode 2002). In the 19th and 20th centuries, these killer whales harried and herded migrating baleen whales into a trap in Twofold Bay, Australia, and got the attention of human whalers living there. After human harpoons finished the prey, whalers left the carcass overnight for the killer whales, who ate just the tongue that they prized. When these conventions were later betrayed, the behavior of the orcas, including ceased cooperation, indicated changes in their interspecies enculturated perspectives. The comradeship, playfulness, excitement, and anger in the orcas in these interspecies engagements shine brightly enough to reach us despite our differences.

Cases like these illustrate that the ideas and perspectives of interspecies enculturated animals are accessible through witnessing their actions and effects, engaging with them socially, empathizing with them, and analyzing experiences of and with them. Human researchers wishing to apply anthropological methods to study, document, and compare dogs' socially learned outlooks in any of the thousands of interspecies cultures that they participate in already have a relational entry point into which they can actually or imaginatively insert themselves: Dogs and humans can connect socially and do so in marvelously varying ways. The emic perspectives that interspecies anthropological engagements can reveal are not just the wonderfully variable ones of the other humans who are passing cultural information back and forth with the other animals; They are those of members of the other species.

What are all those dogs in different interspecies cultures thinking, feeling, expecting, understanding? How do their life-worlds vary, and what do they all have in common? What does dog emics teach us about that greater unit, dog-and-humankind? There are plentiful indicators if we spend some quality anthropological time with them. Thousands of human–canine interspecies cultures, seen from the dogs' perspectives, are awaiting our exploration. We understand a little more about the real world by visiting the perspectives of beings a little different from ourselves. If we want to make dogs' *and* peoples' lives better by any measure, we need to understand their situations and perspectives. In this task, as in others, wagging tails have our backs.

Figure 11.1 (Left to right): the author and Ajax, a basenji who adopted his sister Melanie; Sheboygan, Wisconsin ca. 1974. Photo: Melanie Pomerantz.

References

Alter, Joseph S. 2007. The once and future "apeman": Chimeras, human evolution, and disciplinary coherence. *Current Anthropology* 48(5): 637–652.

Baynes-Rock, Marcus. 2013. Life and death in the multispecies commons. *Social Science Information* 52(2): 210–227.

Baynes-Rock, Marcus. 2015. *Among the bone eaters: Encounters with hyenas in Harar*. University Park: The Pennsylvania State University Press.

Bekoff, Marc. 2007. Wild justice and fair play: Cooperation, forgiveness, and morality in animals. In *The animals reader: The essential classic and contemporary writings*. Linda Kalof and Amy Fitzgerald, eds. Pp. 72–90. Oxford: Berg.

Clode, Danielle. 2002. *Killers in Eden: The true story of killer whales and their remarkable partnership with the whalers of Twofold Bay*. Crows Nest, NSW: Allen & Unwin.

Coren, Stanley. 2004. *How dogs think: Understanding the canine mind*. New York: Free Press.

Coulter, Kendra. 2018. Challenging subjects: Towards ethnographic analyses of animals. *Journal for the Anthropology of North America* 21(1): 58–71.

Dawkins, Richard, and Yan Wong. 2016. *The ancestor's tale: A pilgrimage to the dawn of evolution*. Boston, MA: Mariner Books.

Fitzpatrick, Simon, and Kristin Andrews. 2022. Animal culture and animal welfare. *Philosophy of Science* 89(5): 1104–1113.

Hare, Brian, and Michael Tomasello. 2005. Human-like social skills in dogs? *Trends in Cognitive Science* 9(9): 439–444.

Hartigan, John Jr. 2021. Knowing animals: Multispecies ethnography and the scope of anthropology. *American Anthropologist* 123(4): 846–860.

Horowitz, Alexandra. 2009. *Inside of a dog: What dogs see, smell, and know*. New York: Scribner.

Horowitz, Alexandra. 2021. Considering the "dog" in dog–human interaction. *Frontiers in Veterinary Science* 8: 642821.

Koungoulos, Loukas George, Jane Balme, and Sue O'Connor. 2023. Dingoes, companions in life and death: The significance of archaeological canid burial practices in Australia. *PLoS One* 18(10):e0286576.

Locke, Piers. 2017. Elephants as persons, affective apprenticeship, and fieldwork with nonhuman informants in Nepal. *HAU: Journal of Ethnographic Theory* 7 (1): 353–376.

Lohmann, Roger Ivar. 2006. Field methods. In *Encyclopedia of anthropology*. H. James Birx, ed. Pp. 962–968. Thousand Oaks, CA: SAGE Reference.

Lohmann, Roger Ivar. 2011. Empathetic perception and imagination among the Asabano: Lessons for anthropology. In *The anthropology of empathy: Experiencing the lives of others in Pacific societies*. Douglas Hollan and C. Jason Throop, eds. Pp. 95–116. New York: Berghahn.

Losey, Robert J., Robert P. Wishart, and Jan Peter Laurens Loovers, eds. 2018. *Dogs in the north: Stories of cooperation and co-domestication*. London: Routledge.

Luomala, Katharine. 1960. A history of the binomial classification of the Polynesian dog. *Pacific Science* 14(3): 193–223.

Malinowski, Bronislaw. 1984 [1922]. *Argonauts of the Western Pacific*. Prospect Heights, IL: Waveland.

Mezzenzana, Francesca. 2023. Just like humans: Similarity, difference and empathy towards nonhumans in the Amazonian rainforest. In *Conversations on empathy: Interdisciplinary perspectives on imagination and radical othering*. Francesca Mezzenzana and Daniela Peluso, eds. Pp. 197–215. London: Routledge.

Musharbash, Yasmine. 2018. Yuendumu dog tales. In *Living with animals: Bonds across species*. Natalie Porter, ed. Pp. 17–28. Ithaca, NY: Cornell University Press.

Noske, Barbara. 1993. The animal question in anthropology: A commentary. *Society and Animals* 1(2): 185–190.

Porter, Natalie. 2018. Training dogs to feel good: Embodying well-being in multispecies relations. *Medical Anthropology Quarterly* 33(1): 101–119.

Rose, Deborah Bird. 2011. *Wild dog dreaming: Love and extinction*. Charlottesville: University of Virginia Press.

Russell, Nerissa. 2010. Navigating the human–animal boundary. *Reviews in Anthropology* 39(1): 3–24.

Sanders, Clinton R., and Arnold Arluke. 2007. Speaking for dogs. In *The animals reader: The essential classic and contemporary writings.* Linda Kalof and Amy Fitzgerald, eds. Pp. 63–71. Oxford: Berg.

Savvides, Nikki. 2013. Speaking for dogs: The role of dog biographies in improving canine welfare in Bangkok, Thailand. In *Speaking for animals: Animal autobiographic writing*. Margo de Mello, ed. Pp. 231–243. NewYork: Routledge.

Sundance, Kyra, and Chalcy. 2007. *101 Dog tricks: Step-by-step activities to engage, challenge, and bond with your dog*. Gloucester, MA: Quarry Books.

Thomas, Elizabeth Marshall. 1993. *The hidden life of dogs.* Boston: Houghton Mifflin.

Thomas, Elizabeth Marshall. 2000. *The social life of dogs: The grace of canine company.* New York: Pocket Books.

Thomas, Elizabeth Marshall. 2015. Foreword. In *Among the bone eaters: Encounters with hyenas in Harar.* Marcus Baynes-Rock. Pp. ix–xiii. University Park: The Pennsylvania State University Press.

Titcomb, Margaret. 1969. *Dog and man in the ancient Pacific with special attention to Hawaii.* Honolulu: Bernice P. Bishop Museum.

von Uexküll, Jakob. 1957. A stroll through the worlds of animals and men: A picture book of invisible worlds. In *Instinctive behavior: The development of a modern concept*. Claire H. Schiller, ed. Pp. 5–80. New York: International Universities Press.

Webb, Christine, Becca Franks, Monica Gagliano, and Barbara Smuts. 2023. Untabooing empathy: The benefits of empathic science with nonhuman research participants. In *Conversations on empathy: Interdisciplinary perspectives on imagination and radical othering*. Francesca Mezzenzana and Daniela Peluso, eds. Pp. 216–234. London: Routledge.

Index

For Product Safety Concerns and Information please contact our EU
representative GPSR@taylorandfrancis.com
Taylor & Francis Verlag GmbH, Kaufingerstraße 24, 80331 München, Germany

www.ingramcontent.com/pod-product-compliance
Ingram Content Group UK Ltd.
Pitfield, Milton Keynes, MK11 3LW, UK
UKHW022312250926
475702UK00046B/735

* 9 7 8 1 0 4 1 1 1 1 8 1 8 *